In Catastrophic Times

Critical Climate Change

Series Editors:
Tom Cohen and Claire Colebrook

The era of climate change involves the mutation of systems beyond 20th century anthropomorphic models and has stood, until recently, outside representation or address. Understood in a broad and critical sense, climate change concerns material agencies that impact on biomass and energy, erased borders and microbial invention, geological and nanographic time, and extinction events. The possibility of extinction has always been a latent figure in textual production and archives; but the current sense of depletion, decay, mutation and exhaustion calls for new modes of address, new styles of publishing and authoring, and new formats and speeds of distribution. As the pressures and realignments of this re-arrangement occur, so must the critical languages and conceptual templates, political premises and definitions of 'life.' There is a particular need to publish in timely fashion experimental monographs that redefine the boundaries of disciplinary fields, rhetorical invasions, the interface of conceptual and scientific languages, and geomorphic and geopolitical interventions. Critical Climate Change is oriented, in this general manner, toward the epistemopolitical mutations that correspond to the temporalities of terrestrial mutation.

Isabelle Stengers is professor of philosophy at the Université Libre de Bruxelles. She is trained as a chemist and philosopher, and has authored and co-authored many books on the philosophy of science. In 1993 she received the grand price for philosophy from the Académie Francaise. Her last book published in English is *Thinking with Whitehead: A Free and Wild Creation of Concepts* (Cambridge, MA: Harvard University Press, 2014).

In Catastrophic Times: Resisting the Coming Barbarism

Isabelle Stengers

Translated by
Andrew Goffey

Published by Open Humanities Press in collaboration with meson press 2015
Freely available online at http://dx.medra.org/10.14619/016
http://openhumanitiespress.org/books/titles/in-catastrophic-times

First published in French:
Au temps des catastrophes. Résister à la barbarie qui vient
© Editions LA DÉCOUVERTE, Paris, France, 2009

This is an open access book, licensed under Creative Commons Attribution-NonCommercial-NoDerivatives license. Under this license, authors allow anyone to download, reuse, reprint, modify, distribute, and/or copy their work so long as the authors and source are cited and resulting derivative works are licensed under the same or similar license. No permission is required from the authors or the publisher. Statutory fair use and other rights are in no way affected by the above. Read more about the license at creativecommons.org/licenses/by-nc-nd/4.0/.

ISBN (Print) 978-1-78542-009-2
ISBN (PDF): 978-1-78542-010-8
ISBN (EPUB): 978-1-78542-022-1
DOI: 10.14619/016

Open Humanities Press is an international, scholar-led open access publishing collective whose mission is to make leading works of contemporary critical thought freely available worldwide. This book was published in collaboration with meson press, Hybrid Publishing Lab, Leuphana University of Lüneburg.

Funded by the EU major project Innovation Incubator Lüneburg

Contents

Preface to the English Language Edition

It is 2015 and I find myself in a situation similar to the one I found myself in at the end of 2008, when I was sending the manuscript for this book to the publisher. Was it necessary to make the situation I was discussing "actual" in order to address readers for whom what mattered, what they were in the process of living through was, primarily, the financial crash and its consequences? Or was it necessary to resist the manner in which a history, which is first of all that of a capitalism freed from what had claimed to regulate it, imposes its own temporal horizons?

The necessity of resisting hasn't changed. Governments continue to proclaim their good intentions but "realism" has triumphed. Every measure that would fetter the free dynamics of the market, that is to say, the unalienable right of multinational oil companies and financial speculators to transform every situation, whatever it may be, into a source of profit, will be condemned as "unrealistic." A carbon market, the source of lucrative operations, is perhaps OK, but certainly not the calling into question of extraction rights – we must keep the right to extract and therefore to burn up all the petrol and gas to which we can have access.

Thanks to the increasingly polluting (fracking) or dangerous (deep water) operations for the extraction of "non-conventional" energy sources, the idea of an energy shortage, forcing a transformation of modes of production and consumption, is now behind us. It seems that we have largely sufficient means to produce a degree of warming that would set off an uncontrollable disruption of the climate (runaway climate change). That the earth may then become uninhabitable for species which, like our own, depend on relative climatic stability goes without saying. That it may even, like Venus, become a dead planet is a question to which we will never know the answer.

What I had not foreseen when I was writing *In Catastrophic Times* is that the great "mobilization of America," which everyone in Europe was expecting, would not take place. How many times did we, at that time, hear the comparison with the US entrance into the Second World War. Timid old Europe was doing all it could, but when the Americans finally understood, when they mobilized, then....We could count on the rapid, radical transformation of its economy, with the fervent support of an entire population. As is known, between 2007 and 2011 the percentage of Americans taking climate change seriously collapsed, dropping from 71% to 44%. For all those who were expecting the announcement of more constraining commitments from Copenhagen, there was a rude and painful awakening. Today there is no need to assert, as I did at the time of writing *In Catastrophic Times,* that capitalism—some representatives of which claimed held the solution (so-called green capitalism)—is fundamentally irresponsible. In fact, unregulated capitalism and its allies have refused the role that should have been theirs.[1] It was the route of direct confrontation that was taken, with the determined negation of global warming. "Drill, baby, drill."

Today, the grand campaign to deny the problem has run out of breath a little, but the second phase is being prepared. New voices are making themselves heard, asserting that it is impossible to restrict emissions, which in the meantime have exploded. The only solution is geo-engineering, which will ensure that it is possible to continue to extract and burn, without the temperature rising....

Geo-engineering might only be a dream, or the nightmare of a sorcerer's apprentice. But the radical uncertainty with regard to the catastrophes that it is likely to produce, to say nothing of its effectiveness, won't make the capitalist machine hesitate, because it is incapable of hesitating: it can't do anything other

1 Naomi Klein, *This Changes Everything: Capitalism vs the Climate* (New York: Simon and Schuster, 2014)

than define every situation as a source of profit. At the moment we are at the stage of fiction, but we know that soon this fiction will be proposed, and will try to impose itself, as the only "logical" solution, whether we like it or not. Logical because in effect it respects the demands of those who reject any calling into question of the right to irresponsibility that they have conquered, and confirms that the techno-industrial capitalist path is the only one that is viable. Moreover, it implies the prospect of a mobilization of public finance – but obviously extremely profitably in private hands – and here the example of the US war effort becomes relevant. This solution has an additional advantage, which is that if it should ever work, the war against global warming will never stop. Humanity in its entirety would be taken hostage, constrained to serve masters who will present themselves as its saviors, as those who are protecting it from an invincible enemy who must be kept permanently at a distance. In this way an "infernal alternative" will be fabricated at the planetary scale: either it's us, your saviors, or it's the end of the world.[2]

Today a new word has been created to characterize our situation: our epoch would be the epoch of the anthropocene. One need not be paranoid in order to ask oneself if the success of this word, as much in the media as in the academic world (in a few years the number of conferences and publications on the anthropocene has exploded), doesn't signal a transition from the first phase—of denial—to the second phase—that of the new grand narrative in which Man becomes conscious of the fact that his activities transform the earth at the global scale of geology, and that he must therefore take responsibility for the future of the planet. Of course, many of those who have taken up this word are full of

2 Philippe Pignarre and Isabelle Stengers, *Capitalist Sorcery: Breaking the Spell*, trans. Andrew Goffey (London: Palgrave Macmillan, 2011). The generic formula for the "infernal alternatives," woven since the divorce between capitalism and the great tale of progress has become perceptible, is "you are envisaging resisting this quite unpalatable proposition, but we will show you that if you do the consequences will be worse."

goodwill. But Man here is a troubling abstraction. The moment when this Man will be called on to mobilize in order to "save the planet," with all the technoscientific resources that will be "unhappily necessary," is not far off.

In Catastrophic Times is neither a book of prophecy nor a survival guide. There isn't the slightest guarantee that we will be able to overcome the hold that capitalism has over us (and in this instance, what some have proposed calling "capitalocene," and not anthropocene, will be a geological epoch that is extremely short). Nor do we know how, in the best of cases, we might live in the ruins that it will leave us: the window of opportunity in which, on paper, the measures to take were reasonably clear, is in the process of closing. It wasn't necessary to be a prophet to write, as I have done, that we are more badly equipped than ever for putting to work the solutions defined as necessary. Those– most notably, scientists—who thought that it was enough to sound the alarm neglected the fact that political powers had just handed the rudder over to capitalism and had solemnly renounced any freedom of action.

We do, however, know one thing: even if it is a matter of the death of what we have called a civilization, there are many manners of dying, some being more ugly than others. I belong to a generation that will perhaps be the most hated in human memory, the generation that "knew" but did nothing or did too little (changing our lightbulbs, sorting our rubbish, riding bicycles...). But it is also a generation that will avoid the worst - we will already be dead. I would add that this is the generation that, thirty years ago, participated in, or impotently witnessed, the failure of the encounter between two movements that could, together, perhaps have created the political intelligence necessary to the development of an efficacious culture of struggle[3] - those who denounced the ravaging of nature and those who combated the exploitation

3 This is not knowledge in hindsight. The missed encounter was lived as such. Some voices, like that of Félix Guattari, who, in his *The Three Ecologies*, trans.

of humans. In fact, the manner in which large environmental movements have adhered to the promises of "green" capitalism is enough to retroactively confirm the most somber of suspicions. But the retroactive justification should not erase the memory of a missed opportunity, of a blind division from which the capitalist sirens haven't failed to profit. Capitalism knows how to profit from every opportunity.

What I was afraid of, at the time I wrote *In Catastrophic Times,* was a form of denial on the part of those who saw clearly that the threat of climate change could be an argument mobilized against unproductive conflict as part of the necessary reconciliation between all those of goodwill. Faced with the danger of climate change, a "social peace" could be imposed, and a culpabilizing bureaucratic moralism installed. Hadn't we already started to hear that even the unemployed should learn to reduce their carbon footprint? Today, the fable of a supposedly green capitalism, bringing new, sustainable employment, the agent of peaceful, consensual adaptation of the "systemic" constraints of the climate, is not quite dead. But denying the threat of climate change is no longer necessary in order to denounce this fable. What we are now living is the waking nightmare of a predatory capitalism to which States have handed, in all opacity, the control of the future, laying the burden of the quasi-moral injunction of paying off "their" debts on their own populations and attacking each other before the tribunal of the World Trade Organization (WTO) in reaction to the slightest measure aiming to limit the predation. In short, it is more and more blatantly obvious that the oligarchy of the super-rich has acquired the power to put the world in the service of its interests. Many ecological activists today have become as radically anticapitalist as the militants of the Marxist tradition.

Ian Pindar and Paul Sutton (London: Athlone, 2000) called in vain for the transversality of struggles.

The old suspicions are tenacious, however, as is the attachment to conceptual grand narratives that are perfectly compatible with the mirage of the anthropocene (to wit this call to order from Alain Badiou, for whom ecology is the new opium of the people: "It must be clearly affirmed that humanity is an animal species that attempts to overcome its animality, a natural set that attempts to denaturalise itself."[4]) Whatever the case may be, it is a matter today of at least trying not to let old, reheated hatreds poison the new generation, the generation of activists who, on the ground, are confronting a State rationality that has become the servant pure and simple of the imperatives of growth and competition, and of all those who – often the same – are experimenting with the possibilities of manners of living and cooperating that have been destroyed in the name of progress.

This book was addressed and is still addressed to everyone who is struggling and experimenting today, to everyone who is a true contemporary of what I have dared to call "the intrusion of Gaia," this "nature" that has left behind its traditional role and now has the power to question us all. Formulating this question in a mode that helps them to resist the poisons we have left for them, the grand narratives that have contributed to our blindness, is its only ambition.

4 *Le Grand Soir*, "L'hypothèse communiste," interview with Alain Badiou by Pierre Gaultier, August 2009. http://www.legrandsoir.info/L-hypothese-communiste-interview-d-Alain-Badiou-par-Pierre.html.

Introduction

It is not a question here of demonstrating that the decades to come will be crucial, nor of describing what could happen. What I am attempting instead is of the order of an "intervention," something that we experience during a debate when a participant speaks and presents the situation a little differently, creating a short freezing of time. Subsequently, of course, the debate starts again as if nothing had happened, but some amongst those who were listening will later make it known that they were touched. That is what happened during a debate on Belgian television about global warming, when I suggested that we were "exceptionally ill-equipped to deal with what is in the process of happening." The discovery that such a remark could function as an intervention is the point of departure of this essay.

Intervening demands a certain brevity, because it is not a question of convincing but rather of passing "to whom it may concern" what makes you think, feel, and imagine. But it is also a fairly demanding test, a trajectory where it is easy to slip up, and so which it is important not to try alone. That is why I must give thanks to those who have read this text at one or other stage of its elaboration, and whose criticisms, suggestions, and indeed (above all, even) misunderstandings have guided me and forced me to clarify what I was writing; that is to say, to better understand what this essay demanded.

Thanks first of all to Philippe Pignarre who said "you can" to me from the stage of the first draft, to Didier Demorcy who ceaselessly awakened me to the demands of what I was undertaking, and also to Daniel Tanuro who gave me decisive impetus at a moment when I was seeking the right angle from which to approach my question. Thanks also to Emilie Hache, Olivier Hofman, and Maud Kristen.

Thanks to the members of the Groupe d'études constructivistes, and in particular to Didier Debaise, Daniel de Beer, Marion

Jacot-Descombes, David Jamar, Ladislas Kroitor, Jonathan Philippe, Maria Puig della Bellacasa, and Benedikte Zitouni. Being able to count on the generosity of these researchers, their straight talking, and their practicing of an open and demanding collective intelligence, is a real privilege.

Thanks finally to Bruno Latour whose demanding objections are part of a process that for more than twenty years has testified that agreements between sometimes diverging paths are created thanks to, and not in spite of, divergence.

[1]

Between two Histories

We live in strange times, a little as if we were suspended between two histories, both of which speak of a world become "global." One of them is familiar to us. It has the rhythm of news from the front in the great worldwide competition and has economic growth for its arrow of time. It has the clarity of evidence with regard to what it requires and promotes, but it is marked by a remarkable confusion as to its consequences. The other, by contrast, could be called distinct with regard to what is in the process of happening, but it is obscure with regard to what it requires, the response to give to what is in the process of happening.

Clarity does not signify tranquility. At the moment when I began to write this text, the *subprime* crisis was already shaking the banking world and we were learning about the nonnegligible role played by financial speculation in the brutal price increases of basic foodstuffs. At the moment when I was putting the final touches to this text (mid-October 2008), the financial meltdown was underway, panic on the stock markets had been unleashed, and States, who to that point had been kept out of the court of the powerful, were suddenly called on to try to reestablish order and to save the banks. I do not know what the situation will be when this book reaches its readers. What I do know is that, amplified by the crisis, more and more numerous voices could be heard, explaining with great clarity its mechanisms, the fundamental instability of the arrangements of finance, and the intrinsic danger of what investors had put their trust in. Sure, the explanation comes afterwards and it doesn't allow for prediction. But for the moment, all are unanimous: it will be necessary to regulate, to monitor, indeed to outlaw, certain financial products! The era of financial capitalism, this predator freed from every constraint by the ultraliberalism of the Thatcher-Reagan years, would supposedly have come to an end, the banks having to learn their "real" business again, that of servicing industrial capitalism.

Perhaps an era has come to an end, but only as an episode belonging as such to what I have called the first clear and confused "history." I don't believe that I am kidding myself in thinking that if the calm has returned when this book reaches its readers, the primordial challenge will be to "relaunch economic growth." Tomorrow, like yesterday, we will be called on to accept the sacrifices required by the mobilization of everyone for this growth, and to recognize the imperious necessity of reforms "because the world has changed." The message addressed to all will thus remain unchanged: "We have no choice, we must grit our teeth, accept that times are hard and mobilize for the economic growth outside of which there is no conceivable solution. If 'we' do not do so, others will take advantage of our lack of courage and confidence."

In other words, it may be that the relations between protagonists will have been modified, but it will always be the same clear and confused history. The order-words are clear, but the points of view on the link between these order-words that mobilize and the solutions to the problems that are accumulating—growing social inequality, pollution, poisoning by pesticides, exhaustion of raw materials, ground water depletion, etc.—couldn't be more confused.

That is why *In Catastrophic Times,* written for the most part before the catastrophic financial collapse, has not had to be rewritten. Its point of departure is different. This is because to call into question the capacity of what today is called development to respond to the problems I have cited is to push at an open door. The idea that this type of development, which has growth as its motor, could repair what it has itself contributed to creating is not dead but has lost all obviousness. The intrinsically unsustainable character of this development, which some had announced decades ago, has henceforth become common knowledge; this in turn has created the distinct sense that another history has begun. What we know now is that if we grit our teeth and continue to have confidence in economic growth, we are going, as one says, straight to the wall.

This doesn't signify in the slightest a rupture between the two histories. What they have in common is the necessity of resisting what is leading us straight to the wall. In particular, nothing of what I will write should make us forget the indispensable character of big, popular mobilizations (let us think of the WTO protests in Seattle), which are peerless for awakening the capacities to resist and to put pressure on those who demand our confidence. What makes me write this book doesn't deny this urgency, but responds to the felt necessity of trying to listen to that which insists, obscurely. Certainly there are many things to demand already from the protagonists who are today defining what is possible and what isn't. Whilst struggling against those who are making the evidences of the first history reign, however, it is a matter of learning to inhabit what henceforth we know, of learning what that which is in the process of happening to us obliges us to.

If the, by now common, knowledge that we are heading straight to the wall demands to be inhabited, it is perhaps because its common character doesn't translate the success of a general "becoming consciously aware." It therefore doesn't benefit from the words, partial knowledges, imaginative creations, or multiple convergences that would have had such a success as their fruit, which would have empowered the voices of those who had previously been denounced as bringers of bad news, partisans of a "return to the cave." As in the financial crash, which gave the proof that the financial world was vulnerable in its entirety, it is the "facts" that have spoken, not ideas that have triumphed. Over the last few years one has had to cede to the evidence: what was lived as a rather abstract possibility, the global climatic disorder, has well and truly begun. This (appropriately named) "inconvenient truth" has henceforth imposed itself. The controversy amongst scientists is over, which doesn't signify that the detractors have disappeared but that one is only interested in them as special cases, to be interpreted by their acquaintance with the oil lobby or for their psychosocial particularities (in France, for example, that of being a member of the Academy of Science), which makes them fractious with regard to what disturbs.

Henceforth we "know" and certain observable effects are already forcing climatologists to correct their models, making the most pessimistic of predictions produced by the simulations become increasingly probable. In short, in this new era, we are no longer only dealing with a nature to be "protected" from the damage caused by humans, but also with a nature capable of threatening our modes of thinking and of living for good.

This new situation doesn't signify that the other questions (pollution, inequalities, etc.) move to the background. Instead they find themselves correlated, in a double mode. On the one hand, as I have already underlined, all call into question the perspective of growth, identified with progress, which nonetheless continues to impose itself as the only conceivable horizon. On the other hand, none can be envisaged independently of the others any longer, because each now includes global warming as one of its components. It is indeed a form of globalization that it is a matter of, with the multiple entanglements of the threats to come.

One knows that new messages are already reaching the unfortunate consumer, who was supposed to have confidence in economic growth but who is now equally invited to measure his or her ecological footprint, that is to say, to recognize the irresponsible and selfish character of his or her mode of consumption. One hears it asserted that it will be necessary to "change our way of life." There is an appeal to goodwill at all levels but the disarray of politicians is almost palpable. How is one to maintain the imperative of "freeing economic growth," of "winning" in the grand economic competition, while the future will define this type of growth as irresponsible, even criminal?

Despite this disarray, it is always the very clear logic of what I have called the first history that prevails and continues to accumulate victims. The recent victims of the financial crisis, certainly, but also, and above all, the "ordinary" victims, sacrificed on the altar of growth to the service of which our lives are dedicated. Amongst these victims, there are those who are distant but there are others who are closer. One thinks of those who have drowned in the Mediterranean, who preferred a probable death to the life that they would lead in their country, "behind in the race for growth," and of those who, having arrived amongst us are pursued as "sans-papiers" (illegal immigrants). But it isn't only a matter of "others." Mobilization for growth hits "our" workers, submitted to intolerable imperatives of productivity, like the unemployed, targeted by policies of activation and motivation, called on to prove that they are spending their time looking for work, even forced to accept any type of "job." In my country, the hunting season against the unemployed has been declared open. Public enemy number one is the "cheat," who has succeeded in fabricating a life in the interstices. That this life might be active, producing joy, cooperation, or solidarity, matters very little, or must even be denounced. The unemployed person who is neither ashamed nor desperate must seek to pass unnoticed because they set a bad example, that of demobilization and desertion. Economic war, this war whose victims have no right to be honored but are called on to find every means of returning to the front, requires all of us.

This quasi-stupefying contrast—between what we know and what mobilizes us—had to be recalled so as to dare to put the future that is being prepared under the sign of barbarism. Not the barbarism which, for the Athenians, characterized peoples defined as uncivilized, but that which, produced by the history of which we have been so proud, was named in 1915 by Rosa Luxemburg in a text that she wrote in prison: “Millions of proletarians of all tongues fall upon the field of dishonor, of fratricide, lacerating themselves while the song of the slave is on their lips.”[1]

Luxemburg, a Marxist, affirmed that our future had as its horizon an alternative: “socialism or barbarism.” Nearly a century later, we haven’t learned very much regarding socialism. On the other hand, we already know the sad refrain that will serve as a song on the lips of those who will survive in a world of shame, fratricide, and self-mutilation. This will be: “Unhappily, we have to, we have no choice.” We have already heard this refrain so many times, most notably with regard to the sans-papiers. It signals that what had, to that point, been defined as intolerable, quasi-unthinkable, is in the process of creeping into habits. And we haven’t seen anything yet. It is not for nothing that the catastrophe in New Orleans was such a big shock. What is being announced is nothing other than the possibility of a New Orleans on a global scale—wind power and solar panels for the rich, who will perhaps be able to continue to use their cars thanks to biofuels, but as for the others...

This book is addressed to all of us who are living in suspense. Amongst us there are those who know that they ought to “do something” but are paralyzed by the disproportionate gap

1 Rosa Luxemburg, *The Junius Pamphlet* (Zürich, 1916) https://www.marxists.org/archive/luxemburg/1915/junius/ch01.htm.

between what they are capable of and what is needed. Or they are tempted to think that it is too late, that there is no longer anything to be done, or even prefer to believe that everything will end up sorting itself out, even if they can't imagine how. But there are also those who struggle, who never gave in to the evidence of the first history, and for whom this history, productive of exploitation, of the war of social inequalities that grow unceasingly, already defines barbarism. It is above all not a matter of making the case to them that the coming barbarism is "different," as if Hurricane Katrina was itself a prefiguring of it, and as if their struggles were as a consequence "outmoded." Quite the contrary! If there was barbarism in New Orleans, it was indeed in the response that was made to Katrina: the poor abandoned whilst the rich found shelter. And this response says nothing of the abstraction that some call human selfishness, but rather of that against which they are struggling, of that which, after having promised us progress, demands that we accept the ineluctable character of the sacrifices imposed by global economic competition—growth or death.

If I dare to write nevertheless that they too are "in suspense," it is because what Katrina can figure as a precursor of seems to me to require a type of engagement that, they had judged, it was (strategically) possible to do without. Nothing is more difficult than to accept the necessity of complicating a struggle that is already so uncertain, grappling with an adversary able to profit from any weakness, from any naïve goodwill. I will try to make people feel that it would nevertheless be disastrous to refuse this necessity. In writing this book I am situating myself amongst those who want to be the inheritors of a history of struggles undertaken against the perpetual state of war that capitalism makes rule. It is the question of how to inherit this history today that makes me write.

If we are in suspense, some are already engaged in experiments that try to make the possibility of a future that isn't barbaric, now. Those who have chosen to desert, to flee this "dirty" economic war, but who, in "fleeing, seek a weapon," as Deleuze said.[2] And seeking, here, means, in the first place, creating, creating a life "after economic growth," a life that explores connections with new powers of acting, feeling, imagining, and thinking. Those who are doing this have already chosen to modify their manner of living–effectively but also politically: they do not live in the name of a guilty concern for their "carbon footprint" but experiment with what betraying the role of confident consumer that is assigned to us signifies. That is to say, what it signifies to enter into a struggle against what fabricates this assignation and to learn concretely to reinvent modes of production and of cooperation that escape from the evidences of economic growth and competition. It is to them that this book is dedicated, and more precisely to the possible that they are trying to make exist. It will not for all that be a matter of making myself into their spokesperson, of describing what they are attempting in their place. They are perfectly capable of speaking for themselves, because far from executing a "return to the cave" as some have accused them, they are expert in the use of websites and networks. They have no need of me, but they do need others—like me—to work, with their own means, at creating the sense of what is happening to us.

One should not expect from this book an answer to the question "What is to be done?" because this expectation will be deceived. My trade is words, and words have a power. They can imprison in doctrinal squabbles or aim at the power of order-words—that is why I fear the word degrowth with its threatening arithmetic rationality—but they can also make one think, produce new connections, shake up habits. That is why I honor the invention of the names "Objectors to Growth/Economic Objectors."[3] Words don't

2 Gilles Deleuze and Claire Parnet, *Dialogues* (London: Athlone, 1987). The reference is to George Jackson.

3 "Objecteurs de croissance" after "objecteurs de conscience": a more long-winded translation that would make the point would be to call them "conscientious objectors to economic growth." —Trans.

have the power to answer the question that multiple and entangled threats of what I have called the "second history," on which we are embarked despite ourselves, raises. But they can—and that is what this book will attempt—contribute to formulating this question in a mode that forces us to think about what the possibility of a future that is not barbaric requires.

[2]

The Epoch Has Changed

In the proper sense this book is what one can call an essay. It is well and truly a matter of trying to think, starting from what is in the first place an observation: "the epoch has changed"; that is to say of giving this observation the power to make us think, feel, imagine, and act. But such an attempt is formidable in that the same observation can serve as an argument to prevent us from thinking, and to anesthetize us. In effect, as the space of the effective choices that give a sense to ideas such as politics or democracy has shrunk, those who I will from now on call "our guardians" have had as their task making the population understand that the world has changed. And thus that "reform" today is a pressing obligation. Now, in their case, to reform means to deny what had made people hope, struggle, and create. It means "let's stop dreaming, one must face the facts."

For example, they will say to us let's stop dreaming that political measures can respond to the lightning increase in inequality. Faced with pauperization, one will have to content oneself with measures that are more of the order of public or even private charity. Because it cannot be a question of going back on the evidence that has succeeded in imposing itself over the course of the last thirty years: one cannot interfere with the "laws of the market," nor with the profits of industry. It is thus a matter of learning to adapt, with the sad sigh that kills politics as much as democracy: "sorry, but we have to."

"We have to" is the leitmotif that Philippe Pignarre and I, in *Capitalist Sorcery*,[1] associated with the hold that capitalism has today more than ever, despite the disappearance of any credible reference to progress. Our primary preoccupation was how one is to address capitalism starting from the necessity of resisting this hold. Here I am tackling the same problem, from a complementary point of view. If it is no longer a matter here of echoing the resistance of the antiglobalization – that is to say also, anticapitalist – movement, this is evidently not because it has lost its importance, but because it too is henceforth confronted with a future whose threats have, in a few years, taken a terribly concrete turn. Those who, starry-eyed, put their confidence in the market, in its capacity to triumph over what they can no longer deny but that they call "challenges," have lost all credibility, but evidently that is not enough to give the future the chance not to be barbaric. And the disturbing truth here – when those who are struggling for another world are concerned – is that it is now a matter of learning to become capable of making it exist. That is what the change of epoch consists of, for us all.

To try to think starting from this "fact," that is to say, from that which has, brutally, become commonly evident, is to avoid taking it as an argument ("the epoch has changed, so..."). It is a matter of taking it as a question, and a question that is posed, not in general, but here and now, at a moment when the grand theme of progress has already stopped being convincing. Thus the demonstrations that capitalism gives us an illusion of freedom, that the choices that it allows us are only forced choices, have become

1 Philippe Pignarre and Isabelle Stengers, *Capitalist Sorcery: Breaking the Spell*, trans. Andrew Goffey (London: Palgrave-Macmillan, 2011).

quasi-redundant. One has henceforth to believe in the market to continue to adhere to the fable of the freedom given to each to choose his or her life. It is a matter then of thinking at a moment when the role – that was previously judged crucial – of illusions and false beliefs has lost its importance, without the power of the false choices that are offered to us having been undermined – quite the contrary.

The epoch has changed: fifty years ago, when the grand perspectives on technico-scientific innovation were synonymous with progress, it would have been quasi-inconceivable not to turn with confidence to the scientists and technologists, not to expect from them the solution to problems that concern the development they have been so proud to be the motor of. But here too – even if it is less evident – confidence has also been profoundly shaken. It is not in the least bit ensured that the sciences, such as we know them at least, are equipped to respond to the threats of the future. Rather, with what is called the "knowledge economy,"[2] it is relatively assured that the answers that the scientists will not fail to propose will not allow us to avoid barbarism.

As for States, we know that with a great outburst of enthusiastic resignation, they have given up all of the means that would have allowed them to grasp their responsibilities and have given the globalized free market control of the future of the planet. Even if – it is henceforth the order of the day – they claim to have understood the need to regulate it so as to avoid excesses. That is why I call them our guardians, those who are responsible for us.[3] They

2 I will come back to this question. I restrict myself here to signaling that what here resembles an empty order-word, for use in grand reports bearing on the challenges of the epoch ("our economy is now a knowledge economy…") in fact designates a strong reorientation of public research policy, making partnerships with industry a crucial condition for the financing of research. This amounts to giving industry the power to direct research and to dictate the criteria for its success (most notably by acquiring patents).

3 As English doesn't use the term "responsible" as a noun, "nos responsables" here has been rendered as "our guardians." –Trans.

are not responsible for the future – it would be to give them too much credit to ask them to give an account on this subject. It is for us that they are responsible, for our acceptance of the harsh reality, for our motivation, for our understanding that it would be in vain for us to meddle with the questions that concern us.

If the epoch has changed, one can thus begin by affirming that we are as badly prepared as possible to produce the type of response that, we feel, the situation requires of us. It is not a matter of an observation of impotence, but rather of a point of departure. If there is nothing much to expect on the part of our guardians, those whose concern and responsibility is that we behave in conformity with the virtues of (good) governance, perhaps more interesting is what they have the task of preventing and that they dread. They dread the moment when the rudder will be lost, when people will obstinately pose them questions that they cannot answer, when they will feel that the old refrains no longer work, that people judge them on their answers, that what they thought was stable is slipping away.

Our guardians are predictable enough. If by chance one of them read the lines above and noticed the direction in which I am heading, he will already have shrugged his shoulders: he knows what people are or are not capable of. He knows that the moment that I am evoking, when the rudder goes, will produce nothing other than an unleashing of selfishness, the triumph of demagoguery. I am nothing but an irresponsible elitist who wants to ignore harsh sociological realities.

I don't know what can be understood by "harsh." I know that amongst experimental scientists – where I learned to think – one wouldn't dare to talk in such terms before the corresponding statement "it is thus and not otherwise" had been submitted to multiple tests. Where are the tests here? Where are the active propositions that render it possible and desirable to do differently, that is to say, together for and, above all, with one another? Where are the concrete and collectively negotiated choices? Where are the stories populating imaginations, sharing learning and successes? Where, in schools, are the modes of working together that would create a taste for the demands of cooperation and the experience of the strength of a collective that works to succeed "all together" against the evaluation that separates and judges?

It is necessary to recall all that, that is to say, the manner in which we have been formed, activated, captured, emptied out, not so as to complain about it, but so as to avoid the impotent sigh that would conclude "we can do nothing about it, we are all guilty of being passive" – which is also to say "we must await the hopefully timely measures which, decided elsewhere, will force us to undergo the necessary changes." The sentiment of impotence threatens every one of us but it is maintained by those who present themselves in the name of "hard reality" and say to us "what would you do in our place?"

To call those who govern us our guardians is to affirm that we are not in their place, and that that isn't by chance. And it is also to prevent them, and their allies, from keeping on repeating, with the greatest impunity, what I have called the first history, that of a generalized competition, of a war of all against all, wherein everyone, individual, enterprise, nation, region of the world, has to accept the sacrifices necessary to have the right to survive (to the detriment of their competitors), and obeys the only system "proven to work." Of all the claims to proof that we have been given, that is the most obscene and the most imbecilic. And yet it keeps coming back, again and again, like a refrain, and it asks us to pretend to believe that things will end up sorting themselves out, that, in the place of our guardians, we would do the same thing, and that our own task is limited to insulating our houses, changing our lightbulbs, etc., but also to continue buying cars because growth has to be supported. There isn't anything to discuss here, anything to argue about – that would be to lend this claim some dignity, and to dignify it would be to nourish it. Better instead to renew the virtues of laughter, rudeness, and satire.

Those who I am calling our guardians will protest that to refuse to put oneself in their place, to refuse to argue, to refuse politely to discuss the virtues of the market and its very likely limits, is to refuse debate, that is to say, rational communication, that is, when all is said and done, democracy! Worst, it is to risk panic, the mother of irrationality, opening up the possibility of every kind of demagoguery. Isn't their first role, for the difficult times ahead, to maintain confidence so as to avoid this panic? It is in the name of this sacred task that in the past, officials famously stopped the radioactive dust issuing from Chernobyl at the French borders. But this kind of heroic gesture has since multiplied, to the extent that the unavoidable imperative of having to continue as if nothing was wrong has imposed itself, with no other option than to call on the population to grit their teeth and not lose confidence.

In other words, our guardians are responsible for the management of what one might call a *cold panic*, a panic that is signaled by the fact that openly contradictory messages are accepted: "keep consuming, economic growth depends on it" but "think about your carbon footprint"; "you have to realize that our lifestyles will have to change" but "don't forget that we are engaged in a competition on which our prosperity depends." And this panic is also shared by our guardians. Somewhere they hope that a miracle might save us – which also signifies that only a miracle could save us. It might be a miracle that comes from technology, which would spare us the looming trial, or the miracle of a massive conversion, after some enormous catastrophe. Whilst waiting, they give their blessing to exhortations that aim to make people feel guilty and propose that everyone thinks about doing their own bit, on their own scale – on condition, of course, that only a small minority of us give up driving or become vegetarian, because otherwise that would be quite a blow to economic growth.

I won't go so far as to feel sorry for those who have taken upon themselves the charge of having us behave, but I am convinced that if we succeeded in addressing them in the mode of compassion and not denunciation – as if they were indeed effectively "responsible" for the situation – that address could have a certain efficacy. In any case that is one of the bets of this essay. And the word "essay" finds its full meaning here. It really is a matter of *essaying*, in the pragmatic sense of the term, in the sense that the essay defines what would make it a success. As it happens, if, by speaking of our guardians I have permitted myself to confuse that which, in a democracy, should be distinguished – public officials and politicians – this is not so as to defend a far-reaching conceptual thesis on the definition of the relationship between the State and democratic politics, but to characterize a situation of linguistic confusion that is characteristic of and established under the name of governance. The success of this operation of characterization will be nothing other than what one of those responsible will detest the most – that one refuses to put oneself in their position but that one pities them for being there instead.

Make no mistake: when I come to talk about capitalism and the State in a few pages' time, it won't be a question of definitions that would claim to bring to light the real nature of these protagonists better than previous ones either. I am not amongst those who are searching for a position that allows a permanent "truth" behind which what is now commonly perceived in the mode of a "change of epoch" is to be unveiled. I am trying instead to contribute to the question that opens up when such a change becomes perceptible: "to what does it oblige us?" In this regard I will offer neither a demonstration nor a guarantee, whether founded on history or concepts. I will try to think hand to hand with the question, without giving to the present, in which the pertinence of the responses are at risk, the power to judge the past. But also without giving authoritative power to the responses provided to other questions in the past.

And so essaying this first proposition – addressing ourselves to our guardians in the mode of compassion – doesn't signify that the truth about public powers has at last been unveiled. It is a matter of attempting to *characterize* them in a way that notes that this is where we are, without making this into a destiny, as if the truth of the past was to lead us here, or a scandal, as if they had betrayed their mandate (the idea of such a mandate still supposes the evidence of progress), or even an accident on the way, as if such a route could be defined without any reference to progress.

My approach to the situation that puts "us" into suspense today corresponds to the difference between unveiling and characterizing. To unveil would be to have one's heart set on passing from perplexity to the knowledge that, beyond appearances, judges. On the other hand, to characterize, that is to say, to pose the question of "characters," is to envisage this situation in a pragmatic way: at one and the same time to start out from what we think can be known but without giving to this knowledge the power of a definition. It is what the writer of fiction does when she asks herself what her protagonists are likely to do in the situation she has created. To characterize is to go back to the past starting from the present that poses the question, not so as to deduce this present from the past but so as to give the present its thickness: so as to question the protagonists of a situation from the point of view of what they may become capable of, the manner in which they are likely to respond to this situation. The "we" that this essay has intervene is the we who pose questions of this kind today, who know that the situation is critical but don't know which protagonist's cause to take up.

[3]

The GMO Event

To address those who can be characterized as our guardians today, in the mode of compassion, doesn't mean any kind of sympathy at all, far from it. Rather it is a question of the distance to take, of the determined refusal to share their mode of perception, to allow ourselves to be taken as witnesses for their good intentions. There is nothing much to expect from them, in the sense that there is no point in going in for the torments of disappointment and indignation. But nor is there any point, and this is perhaps more difficult, in engaging in head-on opposition, armed with the evidence of a situation that is confrontational and intelligible only on the basis of this conflict. It's not that the conflict is pointless or "old hat," it is its link with the production of intelligibility that is in question, which threatens to give answers before having learned to formulate questions, of offering certainties before having had the experience of perplexity.

I want to give thanks here to something that has allowed me, amongst others, to live through a learning experience that was crucial for me and without which this essay would not have been written. I'm talking here about the "GMO event," because for me, as for many others, what happened in Europe with the resistance to GMO (genetically modified organisms) marks a before and an after. Not the before and after of a victory. That isn't the case: genetically modified and patented organisms have well and truly invaded the Americas and Asia and, even if they are less frequently associated with their initial claim – responding to the challenge of world hunger – with the production of biomass fuels they found an amazing alternative promise. What made for an event in this epoch that is ours, suspended between two histories, what enabled the European movement of resistance to GMO, to make the possibility of acting rather than undergoing felt, was the discrepancy that was created between the position of those who were in the process of producing more and more concrete, more and more significant knowledges, and the knowledge of those responsible for public order. It may be because of this discrepancy that they were incapable of reconciling opinion with what for them was merely a new agricultural mode of production that illustrated how fruitful the relationship between science and innovation was.

Even the scientific establishment, in general always ready to lay claim to the benefits of an industrial innovation and to shift responsibility for their failings onto others, was shaken up. For example, February 12, 1997 was a terrible moment for French science: the Prime Minister Alain Juppé repudiated the Commission for Biomolecular Engineering by refusing, against their advice, to authorize the launch of three varieties of genetically modified corn. The Commission had a clear conscience. With regard to colza (rapeseed/canola), it certainly restricted itself at first to the "intrinsic" danger of the plant as a product of genetic modification, but gradually started to admit that a flow of genes that induce resistance to herbicides was going to be brought about and could pose a problem. A ban was unimaginable for the Commission but it envisaged possibly setting up a biomonitoring apparatus (in other words, this signified that commercial development would also be an experimental stage, aimed at "better understanding the risk"). But corn didn't pose such problems because it doesn't have parent plants in Europe! The French government had thus done the unforgivable, it had betrayed Science, given way to irrational fears, taken a position in an affair that wasn't its concern, but that of experts.

In fact, the politicians had understood that the situation was out of their control: the scientists were openly divided, public research called seriously into question, militant actions had begun and, in the wake of the so-called mad cow crisis, trust in scientific expertise was at its lowest ebb. But what the politicians hadn't foreseen is that more than ten years later they still wouldn't have succeeded in "calming this down." To their great dismay, and whilst they are subject to enormous pressure on the part of WTO, the United States, industry and its lobbyists, including scientists, European national governments and the European Commission (EC) have not so far succeeded in normalizing the situation. What should have happened without any commotion and without friction would most definitely not.

Worse, and this is where the event is situated for me, the arguments that our guardians were counting on provoked not only responses but above all new connections, producing a genuine dynamic of learning between groups that had hitherto been distinct.

It is important to be able to say "I have learned" from others and give thanks to them. Thus what originally engaged me personally was the ignorant arrogance with which scientists announced a response to the question of world hunger that was "finally scientific." I was also convinced, on the basis of the nuclear precedent, that only the public calling into question of a technology of this kind could produce a knowledge that would be somewhat reliable – in any case more reliable than that of experts who are most frequently in the service of the "feasibility" of an innovation that for them is part of the inevitable ("you can't stop progress!") As it happens I was really quite naïve, because what I didn't know was that what the experts were working with was nothing less than reports prepared by the industry itself, reports that are usually remarkably slim, thanks, we later learned, to sleights of hand testifying to the connivance between industrial consortia and the US administration. And I also did not know that the majority of requests for additional information would come up against "industrial secrecy."

Another point of naïvety was my not knowing that the overwhelming majority of the famous experimental fields, the destruction of which was denounced as irrational, a refusal that science might study the consequences of cultivating GMO crops in an open milieu, were not pursuing this goal in the slightest. It was a matter of agronomic tests prescribed for the approval and thus commercialization of seeds. Another discovery was that for the biologists, it was obvious that "GMO insecticides" would greatly facilitate the appearance of resistant insects, also that Monsanto was organizing a veritable private militia and was encouraging informing on anyone who could be suspected of farming with seeds that it owned, etc.

But the repercussions of the event exceed the case of GMOs alone, leading to the question of what agriculture has become in the hands of seed industries, the seed lines that they select in relation to costly and polluting fertilizers and pesticides, with the resulting double eradication of often more robust traditional seeds and small farmers. And leading also to a veritable "object lesson" bearing on what is on the horizon today with the knowledge economy, to wit the direct piloting of entire sectors of publicly funded research by the private sector. Not only is the primary interest of genetic modification at the end of the day about the appropriation of agriculture through patenting, but it is research itself, in biotechnology and elsewhere, which is henceforth determined by patents, and not just by the possibility of a patent to be had, but by existing patents, which void more and more paths of research of any economic interest. Is it any surprise then, that a heavy and ferocious law of silence weighs on researchers, who are required to stick to the slogan "science at the service of everyone," against what they know to be the case?

If the business of GMO crops was an event it is therefore because there was an effective apprenticeship, producing questions that made both scientific experts and State officials stutter, that sometimes even made politicians think, as if a world of problems that they had never posed was becoming visible to them. What is proper to every event is that it brings the future that will inherit from it into communication with a past narrated differently. At the outset, after having announced the amazing novelty of their creations, the promoters of GMO crops protested that they were in continuity with agricultural practices regarding the matter of seed selection. Today it is this very continuity that is the object of stories that are new or which have hitherto been considered "reactionary," stories that resonate together and open the event up to yet more connections, most notably with those who are learning to renew practices of production that modernization had condemned (the slow food movement, permaculture, networks for the rehabilitation and exchange of traditional seeds, etc.).

Of course, the cry of our guardians has been about "the growth of irrationality," "the fear of change," "ignorance and superstition." But this cry and the noble task that follows from it, that of "reconciling the public with 'its' science," have had little effect. Moreover, the question of the "public" has itself been put in crisis. What do "the people" think? How do they "perceive" a situation? Traditionally, opinion polls responded to this question: one addresses a "representative sample of people" and asks them point-blank about questions that do not necessarily interest them. The business of GMO crops was an occasion when citizen juries demonstrated their capacity to ask good questions, which made the experts stutter – if and only if the apparatus that brings them together effectively allows it. Similarly, some sociologists brought participants in one study into public perceptions together in such a way that the participants felt respected as thinking beings. And the questions and objections that they generated collectively were at the same time both pertinent and very worrying for those who are responsible for us. Thus, besides the question of knowing who would profit from this innovation for which everyone is asked to accept the risks, they posed the question of the tracking of the risks, the famous bio-monitoring that we have been promised: with what resources? How many researchers? Who will pay? Over what period of time? What will happen if things go wrong? Etc.

In fact, the apparently perfectly reasonable demands of these citizens sketch out a landscape that doesn't have much to do with that claimed by the "innovation economy" on which it seems our future depends. For an industrialist they signify having to launch an innovation in a milieu that is actively preoccupied with consequences, that is entitled to detect them, that can set conditions – start small, for example, develop slowly so that one can retrace one's steps – which demands that the promoter of the innovation finances the tracking but doesn't organize it, that insists on all consequences being deployed, on no order-word or promise being taken at face value. A simple contrast: today Monsanto in fact profits directly from the proliferation of "super-weeds" that have developed resistance to its herbicide, Round Up. These superweeds require more than ten or twenty times the usual dose of this product, a product that does not have the innocuousness originally claimed. Lie first, then say it is too late, cover everything with a morality of the inevitable, "you can't stop progress": that is what the freedom to innovate demands.

Today, citizen conferences have become an officially promoted symbol for the participation of the public in innovation, but what has been promoted has also been domesticated. Most of these conferences are organized in such a manner that the participants are guided into giving "constructive" advice, accepting the limits of the questions posed, collaborating just like experts in the production of the label "acceptable": a new type of rating for innovations. The domestication has been all the easier for the fact that apparatuses which induce submission and goodwill

– thinking where and when you are told to think – are easier to put in place than those that induce a capacity to ask worrying questions. The fact of knowing that people can become capable of asking such questions, however, is part of the GMO event. Rather than moaning about this other fact, that it has already "recuperated," it belongs to political struggle to invent the manner in which to make what has thus been learned count.

The GMO event has not been brought to an end. It brought to active, ongoing existence all those whose activation made this event, those who have populated a scene where they weren't expected, where the distribution and the tenor of roles had been arranged in a mode that presupposed their absence. Would biofuels, presented as a miracle solution as much to global warming as to rising fuel prices, have been discredited so quickly without them? Pity the poor EC, which had already promoted this "solution," to the great satisfaction of agricultural industrialists!

One must not go so fast, however. Certainly the GMO event constitutes an exemplary case for the bringing into politics of what was supposed to transcend it: progress resulting from the irresistible advances in science and technology. But it only partially responds to the question of the future. In effect, and contrary to what was the case with GMO, it will not just be a question of refusal. The responsibilities with regard to the accumulation of damages and threats are evident. They do not refer in the first place to those I called our guardians, but to what has defined Earth as a resource to be exploited with impunity. We are not in a court of justice, however, where someone whose responsibility has been established must also answer for what he has done, from whom reparation will be sought. We were able to say "no" to GMO crops, but above all we cannot impose on those who are responsible for the disasters that are looming the task of addressing them. It is up to us to create a manner of responding, for ourselves but also for the innumerable living species that we are dragging into the catastrophe, and, despite this "us" only existing virtually, as summoned by the response to be given.

In order to mark the unprecedented character of this situation, the way in which it messes up habits and judgments, I have decided to name what is coming, which, unlike GMO crops, has neither been willed nor prepared by anyone. What we have to create a response to is *the intrusion of Gaia.*

[4]

The Intrusion of Gaia

It is crucial to emphasize here that naming Gaia and characterizing the looming disasters as an intrusion arises from a pragmatic operation. *To name is not to say what is true but to confer on what is named the power to make us feel and think in the mode that the name calls for.* In this instance it is a matter of resisting the temptation to reduce what makes for an event, what calls us into question, to a simple "problem." But it is also to make the difference between the question that is imposed and the response to create exist. Naming Gaia as "the one who intrudes" is also to characterize her as blind to the damage she causes, in the manner of everything that intrudes. That is why the response to create is not a response to Gaia but a response as much to what provoked her intrusion as to its consequences.

In this essay then, Gaia is neither Earth "in the concrete" and nor is it she who is named and invoked when it is a matter of affirming and of making our connection to this Earth felt, of provoking a sense of belonging where separation has been predominant, and of drawing resources for living, struggling, feeling, and thinking from this belonging.[1] It is a matter here of thinking *intrusion, not belonging*.

But why, one might then object, have recourse to a name that can lend itself to misunderstandings? Why not, one friend asked me, name what intrudes Ouranos or Chronos, those terrible children of the mythological Gaia? The objection must be listened to: if a name is to bring about and not to define – that is, to appropriate – the name can nevertheless not be arbitrary. In this instance I know that choosing the name Gaia is a risk, but it is a risk that I accept, because it is *also* a matter for me of making all of those who might be scandalized by a blind or indifferent Gaia feel and think. I want to maintain the memory that in the twentieth century this name was first linked with a proposition of scientific origin. That is, it is a matter of making felt the necessity of resisting moving on from the temptation of brutally opposing the sciences against the reputedly "nonscientific" knowledges, the necessity of inventing the ways of their coupling, which will be vital if we must learn how to respond to what has already started.

What I am naming Gaia was in effect baptized thus by James Lovelock and Lynn Margulis at the start of the 1970s. They drew their lessons from research that contributed to bringing to light the dense set of relations that scientific disciplines were in the habit of dealing with separately – living things, oceans, the atmosphere, climate, more or less fertile soils. To give a name – Gaia – to this assemblage of relations was to insist on two consequences of what could be learned from this new perspective. That on which we depend, and which has so often been defined as the "given," the globally stable context of our histories and our calculations, is the product of a history of co-evolution, the first artisans and real, continuing authors of which were the innumerable populations of microorganisms. And Gaia, the "living planet" has to be recognized as a "being," and not assimilated into

1 In *Capitalist Sorcery* Philippe Pignarre and I affirmed the political sense of such rituals.

a sum of processes, in the same sense that we recognize that a rat, for example, is a being: it is not just endowed with a history but with its own regime of activity and sensitivity, resulting from the manner in which the processes that constitute it are coupled with one another in multiple and entangled manners, the variation of one having multiple repercussions that affect the others. To question Gaia then is to question something that *holds together* in its own particular manner, and the questions that are addressed to any of its constituent processes can bring into play a sometimes unexpected response involving them all.

Lovelock perhaps went a step too far in affirming that this processual coupling ensured a stability of the type that one attributes to a living organism in good health, the repercussions between processes thus having as their effect the diminishing of the consequences of a variation. Gaia thus seemed to be a good, nurturing mother, whose health was to be protected. Today our understanding of the manner in which Gaia holds together is much less reassuring. The question posed by the growing concentration of so-called greenhouse gases is provoking a cascading set of responses that scientists are only just starting to identify.

Gaia then is thus more than ever well named, because if she was honored in the past it was as the fearsome one, as she who was addressed by peasants, who knew that humans depend on something much greater than them, something that tolerates them, but with a tolerance that is not to be abused. She was from well before the cult of maternal love, which pardons everything. A mother perhaps but an irritable one, who should not be offended. And she was also from before the Greeks conferred on their gods a sense of the just and the unjust, before they attributed to them a particular interest in our destinies. It was a matter instead of *paying attention*, of not offending them, not abusing their tolerance.

Imprudently, a margin of tolerance has been well and truly exceeded: that is what the models are saying more and more precisely, that is what the satellites are observing, and that is what the Inuit people know. And the response that Gaia risks giving might well be without any measure in relation to what we have done, a bit like a shrugging of the shoulder provoked when one is briefly touched by a midge. Gaia is ticklish and that is why she must be named as a being. We are no longer dealing (only) with a wild and threatening nature, nor with a fragile nature to be protected, nor a nature to be mercilessly exploited. The case is new. Gaia, she who intrudes, *asks nothing of us*, not even a response to the question she imposes. Offended,[2] Gaia is indifferent to the question "who is responsible?" and doesn't act as a righter of wrongs – it seems clear that the regions of the earth that will be affected first will be the poorest on the planet, to say nothing of all those living beings that have nothing to do with the affair. This doesn't signify, especially not, the justification of any kind of indifference whatsoever on our part with regard to the threats that hang over the living beings that inhabit the earth with us. It simply isn't Gaia's affair.

That Gaia asks nothing of us translates the specificity of what is in the process of coming, what our thinking must succeed in bringing itself to do: it is a matter of thinking successfully, the event of a unilateral intrusion, which imposes a question without being interested in the response. Because Gaia herself is not threatened, unlike the considerable number of living species who will be swept away with unprecedented speed by the change in their milieu that is on the horizon. Her innumerable co-authors,

2 Offended but not vindictive, because evoking a vindictive Gaia is not just to attribute to her a memory but also an interpretation of what happens in terms of intentionality and responsibility. For the same reason, to speak of the "revenge" of Gaia, as James Lovelock does today, is to mobilize a type of psychology that doesn't seem relevant: one takes revenge against someone, whereas the question of offense is one of a matter of post-factum observation. For example, one says "it seems that this gesture offended her, I wonder why?" Correlatively one doesn't struggle against Gaia. Even speaking of combating global warming is inappropriate. If it is a matter of struggling, it is against what provoked Gaia, not against her response.

the microorganisms, will effectively continue to participate in her regime of existence, that of a living planet. And it is precisely because she is not threatened that she makes the epic versions of human history, in which Man, standing up on his hind legs and learning to decipher the laws of nature, understands that he is the master of his own fate, free of any transcendence, look rather old. Gaia is the name of an unprecedented or forgotten form of transcendence: a transcendence deprived of the noble qualities that would allow it to be invoked as an arbiter, guarantor, or resource; a ticklish assemblage of forces that are indifferent to our reasons and our projects.

The intrusion of this type of transcendence, which I am calling Gaia, makes a major unknown, *which is here to stay,* exist at the heart of our lives. This is perhaps what is most difficult to conceptualize: no future can be foreseen in which she will give back to us the liberty of ignoring her. It is not a matter of a "bad moment that will pass," followed by any kind of happy ending – in the shoddy sense of a "problem solved." We are no longer authorized to forget her. We will have to go on answering for what we are undertaking in the face of an implacable being who is deaf to our justifications. A being who has no spokesperson, or rather, whose spokespersons are exposed to fearsome temptations. We know the old ditty, which generally comes from well-fed experts, accustomed to flying, to the effect that "the problem is, there are too many of us," numbers whose "disappearance" would permit significant energy savings. But if we listen to Lovelock, who has become the prophet of disaster, it would be necessary to reduce the human population to about 500 million people in order to pacify Gaia and live reasonably well in harmony with her. The so-called rational calculations, which result in the conclusion that the only solution is to eradicate the vast majority of humans between now and the end of the century, scarcely dissimulate the delusion of a murderous and obscene abstraction. Gaia does not demand such eradication. She doesn't demand anything.

To name Gaia – that is to say, to associate an assemblage of material processes that demand neither to be protected nor to be loved, and which cannot be moved by the public manifestation of our remorse, with the intrusion of a form of transcendence into our history – ought not especially to shock most scientists. They themselves are in the habit of giving names to what they recognize has the power to make them think and imagine – and this is the very sense of the transcendence that I associate with Gaia. Those who have set up camp in the position of the guardians of reason and progress will certainly scream about irrationality. They will denounce a panicky regression that would make us forget the "heritage of the Enlightenment," the grand narrative of human emancipation shaking off the yoke of transcendences. Their role has already been assigned. After having contributed to skepticism with regard to climate change (think of Claude Allègre[3]), they will devote all their energy to reminding an always credulous public opinion that it must not be diverted, that it must believe in the destiny of Man and in his capacity to triumph in the face of every challenge. Concretely, this signifies the duty to believe in science, the brains of humanity, and in technology, in the service of progress. Provoking their yelling is something that neither amuses nor scares me.

The operation of naming is therefore not in the least bit anti-scientific. On the other hand, it may make scientists think, and prevent them from appropriating the question imposed by the intrusion of Gaia. Climate scientists, glaciologists, chemists, and others have done their work and they have also succeeded in making the alarm bells ring despite all the attempts to stifle them, imposing an "inconvenient truth" despite all the accusations that have been leveled against them, of having mixed up science and politics, or of being jealous of the successes of their colleagues, whose work has succeeded in changing the world where

3 French politician and scientist, minister of education under Lionel Jospin, and visible climate change skeptic. –Trans.

theirs has been limited to describing it, or even of presenting as "proven" something that is only hypothetical. They have been able to resist because they knew that time counted, and that it wasn't them, but that to which they were addressing themselves that in fact mixed up scientific and political questions, or, more precisely, aimed at substituting itself for politics and imposing its imperatives on the entire planet. To name Gaia is finally to help scientists resist a new threat, one which this time would fabricate the worst of confusions between science and politics: that one ask them how to respond, that one trust in them to define what it is appropriate to do.

Moreover, that is what is in the process of happening, but with other types of "scientists." Nowadays it is economists who have become active, and in a way which guarantees that like many unwanted effects, the climate question will be envisaged from the point of view of strategies that are plausible, that is to say, are likely to make it a new source of profit. Even if this means being resigned – in the name of economic laws (which are harsh, they will affirm, but which are laws, after all) – to a planetary New Orleans. Even if it means that zones on the planet that are defined as profitable must, at all scales – from the neighborhood to the continent – protect themselves by every means necessary from the mass of those who will doubtless be opposed to the famous "we cannot take care of all the woes of the world." In short, even if the succession of "sorry, but we musts" establishes, completely, and openly deployed, the barbarism that is already in the process of penetrating our world.

Economists and other candidates for the production of global responses based on "science" only exist for me as a power to harm. Their authority only exists to the extent that the world, our world, remains what it is – that is to say, destined for barbarism. Their laws suppose, above all, that we stay in our places, keep the roles assigned to us, that we have the blind self-interest and congenital incapacity to think and cooperate that makes an all azimuths economic war the only conceivable horizon. It would be completely pointless to name Gaia if it was just a matter of combating them. But it is a matter of combating what gives them their authority. Of that against which the cry "another world is possible!" was raised.

This cry really hasn't lost any of its topicality. Because that against which it was raised – capitalism, the capitalism of Marx, of course, not of American economists – is already busying itself concocting its own responses to the question imposed on us, responses that lead straight to barbarism. This is to say that the struggle assumes an unprecedented urgency but that those who are engaged in this struggle must also face a test that they didn't really need, which, in the name of that urgency they might be tempted to abstract out. To name Gaia *is to name the necessity of resisting this temptation*, the necessity of starting out from the acceptance of this testing challenge: *we do not have any choice, because she will not wait.*

Do not ask me to sketch what other world may be able to come to terms, or compose, with Gaia. The response doesn't belong to us, that is to those who have both provoked her intrusion and now decipher it through data, models, and simulations. Naming Gaia is naming a question, but emphatically not defining the terms of the answer, as such a definition would give us, us again, always us, the first and last word. Learning to compose will need many names, not a global one, the voices of many peoples, knowledges, and earthly practices. It belongs to a process of multifold creation, the terrible difficulty of which it would be foolish and dangerous to underestimate *but which it would be suicidal to think of as impossible.* There will be no response other than the barbaric if we do not learn to couple together multiple, divergent struggles and engagements in this process of creation, as hesitant and stammering as it may be.

[5]

Capitalism

I have spoken of those who are responsible for us, those who have assumed the role of our guardians and present themselves as such, even whilst they are in a state of frozen panic. On the other hand, what Marx called capitalism doesn't experience a panic of this kind, even whilst the type of development it is responsible for is called into question by the intrusion of Gaia. And it experiences neither panic nor even hesitation because, quite simply, *it is not equipped for that*. That in any case is why one can inscribe oneself in Marx's heritage without for all that being "Marxist." Those who say to us "Marx is history," with an obscene, satisfied little smile, generally avoid saying to us why capitalism such as Marx named it is no longer a problem. They only imply that it is invincible. Today those who talk about the vanity of struggling against capitalism are de facto saying "barbarism is our destiny."

If we need, now more than ever, perhaps, the manner in which Marx outlined capitalism – even if it means "characterizing" it where he proposed to define it – this is first of all so as not to entertain the hope that, necessity being the law, "they" will end up doubting, understanding that it is the future which is at stake here, that of their children as well as ours. That is to say also so as not to waste our time becoming indignant, denouncing, finally only to draw the darkest of conclusions about the flaws of the species, which in the end would only be getting what it deserves. What Marx named capitalism doesn't speak to us about humans, it doesn't translate their greed, their self-interest, or their inability to pose questions about the future.

Of course – and this is the very sense of Marx's characterization of capitalism – businessmen, as individuals, are like everyone else. It is not impossible that in the 1980s, some may have believed in the "citizen enterprise," to which it was a matter of reconciling the French people. Are they the same bosses or different ones who now remind us that at a time of outsourcing and mergers the only business of the enterprise is to make money? The question is insignificant: the conjuncture has changed. Similarly today certain amongst them are perhaps terribly disturbed, whilst others place their trust in the market, whose capacity for adaptation and innovation should respond to the problem posed by the intrusion of Gaia. When it is a matter of capitalism, individual psychology is completely irrelevant. Capitalism must be understood instead as a mode of functioning, a machine, which fabricates its own necessity, its own actors, in every conjuncture, and destroys those who haven't been able to saddle up for the new opportunities.

In their own way this is what is recognized by the servile or divinatory economists who talk about the laws of the market that impose themselves whatever our projects and futile hopes might be. Capitalism does, in effect, have something transcendent about it, but not in the sense of the laws of nature. Nor in the sense that I have associated with Gaia either, which is most certainly implacable, but in a mode that I would call properly materialist, translating the untameable character of assemblages that couple together those material processes on whose stability what has been called development thought it could count. Capitalism's mode of transcendence *is not implacable, just radically irresponsible*, incapable of answering for anything. And it has nothing to do with the materialism that people of faith often associate with it. In contrast to Gaia, one ought to associate it instead with a power of a (maleficent) "spiritual" type,[1] a power that captures, segments and redefines always more and more dimensions of what makes up our reality, our lives, our practices, in its service.

That I have been led to characterize both the assemblage of coupled material processes that I named Gaia and the regime of economic functioning that Marx named capitalism by a mode of transcendence highlights the particularity of our epoch, that is to say, the global character of the questions to which they oblige us in both cases. The contemporaneity of these two modes of transcendence is evidently no accident: the brutality of the intrusion of Gaia corresponds to the brutality of what has provoked her, that of a development that is blind to its consequences, or which, more precisely, only takes its consequences into account from the point of view of the new sources of profit they can bring about. But the questions of contemporaneity they pose don't imply any confusion within the responses. Struggling against Gaia makes no sense – it is a matter of learning to compose with her. Composing with capitalism makes no sense – it is a matter of struggling against its stranglehold.

You will have understood that to trust in capitalism as it presents itself today, as the "best friend of the earth," as "green," concerned about protection and sustainability, would be to commit the same kind of error as the frog in the fable, who agrees to carry a scorpion on his back across the river. If the scorpion stung him, wouldn't they both drown? And yet the scorpion stings him, right in the middle of the river. With his last breath the frog murmured "why?" to which the scorpion, just before sinking,

1 What Philippe Pignarre and I have associated with the power of the sorcerer to cast spells.

responded "it's in my nature, I couldn't help doing it." It is in the nature of capitalism to exploit opportunities: *it cannot help doing it.*

The logic of capitalist functioning cannot do anything other than identify the intrusion of Gaia with the appearance of a new field of opportunity. Questioning the (bronze-cast) laws of free exchange is something, then, that it cannot envisage. On the other hand, carbon quotas are welcome, permitting as they already do highly fruitful financial operations. Correlatively, the GMO event clearly translates what must, from the point of view of this logic, be avoided, what those who are responsible for us have taken it upon themselves to avoid, and which will have to be so all the more when the effects of Gaia's intrusion become catastrophic: the production of collective capacities to meddle with questions that concern the common future. Discussing details of a solution may be tolerated, but meddling with the manner in which questions are formulated will not.

Governance is well named. It describes well the destruction of what is implied by a collective responsibility with regard to the future, that is to say, politics. With governance, it is no longer a matter of politics but of management, and, in the first place, the management of a population that must not meddle with what concerns it. In the case of GMO crops, our guardians failed in the task they were allotted, from the point of view of the distribution of what capitalism makes the State do and what the State leaves capitalism free to do. They could not get people to accept that GMO crops constituted, if not a benefit for humanity, at least a fact that it was impossible to resist. They were not able to leave capitalism free to do what it had the opportunity to do thanks to GMOs – complete the redefinition of agriculture by submitting it to patent law. Or not without snags, friction or noise, at least. Capitalism doesn't like noise.

But we must not go too quickly and constitute the refusal of GMOs as a model for the unexpected resistance with which the operative logic called capitalism collided. Not just because, of course, GMOs are now almost everywhere – the success is the "almost" – but above all because this refusal benefited from the effect of surprise. GMOs were supposed to happen without too much friction, in the name of the progress that the holy alliance between scientific research and human progress brings. Industrial consortia and their scientific allies noted, to their great consternation, that it no longer worked as a charm, that the reference to progress had lost (part of) its power. But one can think that the lesson has been learned and that in future, the progress argument – which turned out to be unable to create a consensus – will be replaced by the kind of well-concocted montages of what Philippe Pignarre and I, in *Capitalist Sorcery*, called "infernal alternatives."

Everyone is now familiar with what these alternatives produce: "you refuse to accept a reduction in living standards and are calling for a raise? Then business will locate elsewhere...."; "you refuse to accept unbearable levels of work? Then there are plenty of others who will happily take your place...." Every situation in which infernal alternatives are imposed is the "the fruit of patient processes of fabrication at a very small scale, of careful experiment, because it is always a question of capturing without creating too much alarm, or by creating false alarms."[2] What presents itself as a logical consequence (then...) has been fabricated by multiple processes of so-called rational reorganization that in the first place aimed at sapping or capturing the capacities for thinking and resisting of those who were apt to do so. That is why infernal alternatives first concerned the world of labor – questions of retirement, flexibility, salaries, the organization of work, etc. Today, the rhetoric announcing that it is impossible or suicidal to refuse what one doesn't want has become autonomous. Thus, we are told that to struggle against the exorbitant price of patented medicine, even if only for the poor, is to

2 Pignarre and Stengers, *Capitalist Sorcery*, 28.

condemn the research that will produce tomorrow's medicine. The rhetoric, as repeated at the level of the State, is now sufficient to freeze a situation.

It is such a rhetoric that was used in face of the unexpected refusal of GMOs. Alternatives with an infernal mission were quickly cobbled together – for example "if you refuse GMOs, there will be a brain drain," or "you will slow us down in the great economic race," or "you won't get the second-generation GMOs, which will be really beneficial." But it was too late and not at all convincing, because the proposition "GMO" didn't crown a montage that would authorize the infernal argument "if you refuse GMOs, the result will be worse." We can, however, foresee the proliferation of more convincing alternatives. Biofuels prefigure this type of alternative – either a major energy crisis or the forfeiting of a considerable share of productive land. Food riots risk complicating the argument, but the capitalist scorpion can't help it, opportunities must be taken advantage of – in this instance, speculating on – and thus accelerating – the price rise of staple foodstuffs.

Numerous alternatives of the "it is either that, or you will contribute to the climate catastrophe" must therefore be expected. Whilst the intrusion of Gaia won't make capitalism think or hesitate, because capitalism doesn't think or hesitate, such predictable alternatives can make those who have been able to resist being bewitched by capitalism hesitate. They have every reason to dread that in the face of climate threat a mobilization that will claim to transcend conflicts will be proposed. I anticipate and equally dread such appeals to sacred unity and the accusations of betrayal that automatically accompany them. But what I also dread is that this might incite those who resist only to pay lip service to the idea that global warming is effectively a new problem, following it immediately with the demonstration that this problem, like all the others, should be blamed on capitalism, and then by the conclusion that we must therefore maintain our heading, without allowing ourselves to be troubled by a truth that *must not upset* the prospects for the struggle. Those who, like me, take note that it is a matter of learning *from now* on what a response to the intrusion of Gaia that is not barbaric requires, and insist on the necessity of new practices of struggle, are considered naïve.

Such practices (it must be repeated again and again) should not be thought of as the successor of social struggles but in terms of their coupling. But the challenge of such a coupling is in effect formidable because it undoubtedly means complicating the logic of strategic priorities that has till now predominated. What has to be given up, at the moment of greatest need, really is what has most often served as the rudder for struggle – the difference to be made between what this struggle demands and what will become possible afterwards, if capitalism is finally defeated. Naming Gaia, she who intrudes, signifies that *there is no afterwards*. It is a matter of learning to respond now, and notably of creating cooperative practices and relays with those whom Gaia's intrusion has already made think, imagine, and act. With the conscientious objectors to economic growth, for example, and the inventors of "slow" movements, who refuse what capitalism presents as rationalization and who seek to reclaim what it has destroyed. Alliances will be required, and certainly demanding ones, but the demand should not be that of judges who verify that what they are dealing with really is entitled to claim that it constitutes a force of opposition to capitalism, indeed, which even consults the codex in which Marx has already listed untrustworthy allies. Because these new actors will not, almost by definition, have the required legitimacy.

That this is a matter for confusion I can easily understand. But what I dread is that this confusion might be translated into a defensive reaction, into an "I am aware but all the same" that paralyzes and anesthetizes. And I dread just as much that the possible alliance with these new actors be based on tolerance, on the indulgence that adults who "know" reserve for naïve children – adults who will continue to think between themselves whilst encouraging the goodwill of young idealists. It is a matter of taking note of the fact that Gaia's intrusion questions the theories that armed this "adult" knowledge, which was supposed to provide a compass for struggles, to allow the direction to be maintained despite all the false pretenders, illusions, and chimera that the Great Illusionist systematically produces. To throw away such a compass at the moment when it is a matter of confronting a capitalism that is more powerful than ever might appear to be the height of irresponsibility. Naming Gaia is accepting to think with this fact: *there is no choice.*

This "there is no choice" is one that materialists ought to be able to accept. But here it is a matter of not just "accepting because there are no means of doing otherwise." It is a matter of being obliged to think by what happens. And perhaps the test will demand the abandoning, without any nostalgia, of the heritage of a nineteenth century dazzled by the progress of science and technology, cutting the link then established between emancipation and what I would call an "epic" version of materialism, a version that tends to substitute the tale of a conquest of nature by human labor for the fable of Man "created to have dominion over the earth." It is a seductive conceptual trick but one that bets on an earth available for this dominion or conquest. Naming Gaia is therefore to abandon the link between emancipation and epic conquest, indeed even between emancipation and most of the significations that, since the nineteenth century, have been attached to what was baptized "progress." Struggle there must be, but it doesn't have, can no longer have, the advent of a humanity finally liberated from all transcendence as its aim. *We will always have to reckon with Gaia,* to learn, like peoples of old, not to offend her.

People will perhaps say that my sketch is a simplification or a caricature. Certainly, and it is not a matter here of knowing what is in Marx's texts and what isn't. If I caricature, it is in order to characterize the test, the difficulty for us of thinking that the challenge of Gaia's intrusion cannot be reduced to a "bad moment that will pass" together with capitalism, which is responsible for it. Indifferent to human reason, blind to the greatness of what we call emancipation, this intrusion puts all those whom it challenges on an equal footing because no knowledge can claim any privilege with regards to the response to bring to it. Not that what we know is henceforth null and void. Definitely not. It is the consequences of what we know that stammer, that is to say, the set of "and so..." that adults and judges fabricate.

Accepting the challenge doesn't signify, for me, calling into question the notion of emancipation itself, the idea that there are childish dependences that we must learn to rid ourselves of. But the point of view changes a little. If there is a childish dependence, it is above all *ours*, our dependence on the confidence that we placed in the epic fable of Progress, in its multiple and apparently discordant versions, all of which nevertheless converge in blind judgments about other peoples (to be liberated, modernized, educated, etc.). If there must be emancipation, it will have to be carried out against what has allowed us to believe we can define a heading that would provide a direction for the progress of the entirety of humanity, that is to say, against the hold of the clandestine form of transcendence that has seized us. There are many names for this transcendence, but I will characterize it here by the strange right that has prevailed in its name, a right that would have frightened all the peoples who knew how to honor divinities such as Gaia, because it is a matter of the *right not to pay attention.*

[6]

Not Paying Attention!

The need to pay attention is, apparently, common knowledge. We know how to pay attention to all sorts of things, and even those who are attached most ferociously to the virtues of Western rationality will not refuse this knowledge to peoples whom they disqualify as superstitious. Furthermore, even animals on the lookout testify to this capacity....

And yet we can also say that once it is a matter of what one calls "development" or "growth," the injunction is above all to not pay attention. Growth is a matter of what presides over everything else, including – we are ordered to think – the possibility of compensating for all the damage that is its price. In other words, whilst we have more and more means for foreseeing and measuring this damage, the same blindness that we attribute to civilizations in the past (who destroyed the environment on which they depended) is demanded of us. They may not have understood what they were doing, and they did it only locally. We know that we are destroying to the point of scarcity resources constituted over the course of millions of years of terrestrial history (much longer for aquifers).

What we have been ordered to forget is not the capacity to pay attention, but the *art* of paying attention. If there is an art, and not just a capacity, this is because it is a matter of learning and cultivating, that is to say, making ourselves pay attention.[1] Making in the sense that attention here is not related to that which is defined as a priori worthy of attention, but as something that creates an obligation to imagine, to check, to envisage, consequences that bring into play connections between what we are in the habit of keeping separate. In short, making ourselves pay attention in the sense that attention requires knowing how to resist the temptation to separate what must be taken into account and what may be neglected.

The art of paying attention is far from having been rehabilitated by the precautionary principle, although the protests of industrialists and their scientific allies give us a foretaste of what that rehabilitation would signify. When one hears the protestations that continue today against this unfortunate principle, one can only be seized by a certain fright, as much because of the contempt they express in relation to a population defined as being scared of everything and nothing, calling for zero risk, as because of the feeling of legitimacy of those protesting, those brains of humanity who are charged with the task of guiding the human flock towards progress. Because this principle is apparently perfectly reasonable: it is restricted to affirming that in order to take into account a serious and/or irreversible risk to health or the environment, it is not necessary that that this risk be scientifically proven. In other words, what has provoked so much protest is limited to stating that even if the risk is not proven, one is supposed to pay attention.

1 It is not easy to directly translate the sense of these passages here and capture their broader resonances. What I have translated as "making ourselves pay attention" is, in the original French, "faire attention" – where "faire" means "to make" or "to do." "Faites attention" also means something like "look out" or "be careful." –Trans.

Health and environmental catastrophes have been necessary for the public powers in Europe to finally be constrained to acknowledge that a precautionary principle is well founded. That some renowned scientists have been able to cry out "betrayal" despite such catastrophes casts a very strange and raw light on the situation that it is the ambition of this principle to reform: a paradoxical situation, as the necessity of paying attention where there are doubts, what one would require of a "good father," what one teaches children, is defined here as the enemy of Progress.

Yet what made those scientists cry out was rather timid, because the precautionary principle respects the precoded stage on which it intervenes, a stage on which the task of judging the value of an industrial innovation is entrusted solely to its encounter with the market, and in which public powers only have the right to place certain conditions on this encounter. The principle is limited to extending this right a little, but doesn't modify the logic of the scenario at all. Evaluation continues to belong to the market and therefore only involves the criteria that the market accepts. As for the conditions in which the principle is applied, they are extremely restrictive. Not only must risks bear on health or the environment, and therefore not concern, for example, the social catastrophes that an innovation can provoke, but the principle indicates that the measures that respond to the taking into account of the risk must be "proportionate." One might think that proportionality would bear on any evaluation of the benefits of a techno-industrial innovation for the general interest, since that is what is in play with the risk. But no, what proportionality puts on stage is concern for the damages that the measures will entail for those who benefit from the sacred right of the entrepreneur, the sacred right of bringing things to market, of making them circulate.

So, can Monsanto's right as an entrepreneur be questioned, on the pretext that GMOs clearly risk accelerating the proliferation of insects that are resistant to the pesticide loaded into plants? Certainly not. One is limited to enacting rules that aim to reduce the probability that such insects will appear, and to hoping that the agriculturalists concerned will obey these rules, which will permanently complicate their lives and reduce the profits they were banking on. Since prohibiting Monsanto's GMOs would be a "disproportionate" measure, no other choice can be envisaged. As for the socio-economic consequences of GMOs – there is no place for them. Ruining Indian peasant smallholders is not a serious or irreversible risk, even if they commit suicide. It's the price, harsh but necessary, of the modernization of agriculture.

It will be said that it is entrepreneurial freedom that is at stake. And every entrepreneur will repeat the refrain: risk is the price of progress (today: of competitiveness). But here is where we must slow down and pay attention. To agree to identify Monsanto with the entrepreneur whose heroic stance it claims, that of one who accepts the possibility of failure with a valiant heart, that of the Promethean man who is incessantly exploring what could become possible, is to allow oneself to be trapped by one of those dramatic stagings that are the trademark of master thinkers relating the intrusion of Gaia to the audacity of Man, who has dared to challenge the order of things. From which the consequences cascade, pushing us up against the wall: have confidence in the genius of humanity, or curse it and repent. Well, well! But hasn't capitalism been forgotten?

The heroic pose struck by Monsanto and others like it is misplaced. Because when it is a matter of their own investments, it is *security* that they demand: only the market, a veritable judgment of God, can be called on to put them at risk, not the question of consequences. That this judgment of God is itself rigged goes without saying. On the other hand, that these so-called entrepreneurs, who assume the passion for what may be possible, can demand that the question of possible consequences not constitute an argument entitled to put them at risk is what matters to me here.

In order to separate those with whom we are dealing from this story about creative and audacious entrepreneurs, which they claim to be a part of, commanding us to choose between the adventure of humanity and fearful renunciation, I will call them Entrepreneurs, the capital letter signifying – as will be the case later, with Science – that it is a matter of a façade that dissimulates a change of nature. We will not say that the Entrepreneur has a (Promethean) confidence in progress, which "can mend whatever damage it may have occasioned," a confidence that compels us all to face the grandeur of Man's vocation and his future (written in the stars?) What the double scandal of the GMO event and the precautionary principle for our Entrepreneurs and their allies teaches us, is that it is not a matter of confidence. *It really is a matter of a demand.* Correlatively, relearning the art of paying attention has nothing to do with a sort of moral imperative, a call for respect or for a prudence that we might have lost. *It is not a matter of "us," but of business, which the Entrepreneur requires us not to meddle with.*

When Marx characterized capitalism, the big question was "who produces wealth?" hence the preponderance of the figure of the Exploiter, this bloodsucker who parasitizes the living power of human labor. Evidently this question has lost nothing of its currency, but another figure might be added, without any rivalry, to this first, corresponding to the injunction not to pay attention, including even when barbarism threatens. This figure is the Entrepreneur, he for whom everything is an opportunity, or rather, he who *demands the freedom to be able to transform everything into an opportunity* – for new profits, including what calls the common future into question. "This could be dangerous" is something that an individual chief executive officer (CEO) might understand, but not the operative logic of capitalism, which will eventually condemn whoever recoils in the face of an entrepreneurial possibility.

With the figure of the Entrepreneur come two others, because the Entrepreneur demands, and his demand must be heard and satisfied. These two figures are the State and Science. One could perhaps associate the moment when one can really talk about capitalism with the moment when an Entrepreneur can count on a State that recognizes the legitimacy of his demand, *that of a "riskless" definition of the risk of innovation.* When an industrialist says, with the tears of the sacred in his voice, "the market will judge," he is celebrating the conquest of this power. He doesn't have to answer for the consequences (which are possibly highly undesirable) of what is put on the market, except if these contravene a regulation explicitly formulated by the State, a scientifically motivated regulation that responds to the imperative of proportionality. As for Science, which has been accorded a general authority for all terrains about the definition of the risks that must be taken into account, it has little to do with the sciences. One will not be astonished that the experts who play this game know that their opinions will not be plausible unless they are as balanced as possible, that is to say, give all due weight to the legitimacy of the innovator who has "made the investment."

What is this Science, which intervenes here as the third thief, an arbiter tolerated by the Entrepreneur with regard to his right to innovate, that is to say too, with regard to the right that he recognizes (albeit constrained and forced) the State has to prohibit or regulate? If I have given it a capital letter, it is to distinguish it from scientific practices. And that not so as to exempt practitioners of any responsibility, to oppose experts (in the service of power) with (disinterested) researchers, but because with the coupling together of Entrepreneur, State, and Science, we are very close to the gilded legend that prevails whenever it is a question of the "irresistible rise to power of the West." This legend, in effect, stages the decisive alliance between scientific rationality, the mother of progress of all knowledge, the State, finally free of the archaic sources of legitimacy that prevented this rationality from developing, and the industrial growth that translates what Marxists have called the development of the forces of production into an at last unbounded principle of action. It is from the grip of this legend that it is a matter of escaping, of course. And if the art of paying attention must be reclaimed, what matters is to begin by paying attention to *the manner* in which we are capable of escaping it.

Here again it will not be a matter of defining the truth of the State or of Science, of rewriting the "real story" behind the legend, but of activating questions that arise first of all from the moment in which we are living, from what it forces us to think and also from what it asks us to be wary of. What it is a matter of being wary of are the simplifications that would still ratify a story of progress, including the one that enables us to see the truth of what we are facing. Whether this truth makes capitalism the only real protagonist, the relative autonomy of the authors being largely illusory, or makes the three protagonists the three heads of the same monster, which it behooves the interpreter to name, what is missing is the question, which has become crucial today, of knowing what might or might not be a resource for the task of learning once again the art of paying attention.

[7]

A Story of Three Thieves

I have written a great deal about the sciences, and notably against their identification with an undertaking that would be neutral, objective, and finally rational. It was a matter not of attacking scientific practices but of defending them against an image of authority that is foreign to what makes for their fecundity and relative reliability.[1] I will restrict myself here to emphasizing that whenever it is a question of scientific research, the definition of what "must" be taken into account never imposes itself in a general manner, but translates *the event of an achievement* that opens up a new field of questions and possibilities to those that it concerns. Science, with a capital 'S,' is a stranger to this type of event, and it participates directly in the prohibition that bears on paying attention.

"It's unproven, it's unproven!" How many times have experts made this obscene refrain ring out? A refrain whose authority,

1 See on this subject Isabelle Stengers, *The Invention of Modern Science,* trans. Daniel W. Smith (Minneapolis: Minnesota University Press, 2000) and *La Vierge et le neutrino* (Paris: Les Empêcheurs de penser en rond, 2006).

it really must be emphasized, is not called into question by the precautionary principle, because this talks of what is "not yet" proven. With Science, it is no longer a question of proof as an achievement, as something of an event. Proof is *what one is entitled to demand*, where a question, an objection, a worrying proposition, surfaces. The primary role of the refrain "it's unproven" is to shut up, to separate out what is reputedly objective or rational from that which will be rejected as subjective, or illusory, or as the manifestation of irrational attachment to ways of life that unfortunately progress condemns. This role, accepted by many scientists, dishonors those who endorse it much more intimately even than their participation in the development of weapons. Because it transforms the event that genuine proof constitutes, the rare achievement the possibility of which puts researchers into tension, forcing them to think, to object, to create, into an all-terrain imperative.

But this dishonor goes back a long way. I have tried to characterize the practical novelty effectively associated with the experimental proof with the realization that certain facts – those that will be called experimental – may be recognized as having the power to testify to the manner in which they must be interpreted. That is the achievement, passionately staged and verified in laboratories, which makes experimental scientists, those who understand what it means to dance in the laboratory when it works, think, imagine, bustle about, or object. But Galileo, who discovered that such an achievement was possible, hastened to generalize it, that is to say, to transform the event (to succeed in producing a type of fact that "proves") into the reward of an at last rational method (to yield to the facts). Thus he was able to oppose the new scientific reason, which only accepts the authority of facts, to all those who took sides on undecidable questions, who gave power to their convictions or prejudices. This staging was without a doubt *one of the most successful propaganda operations in human history,* as it has been repeated and ratified even by the philosophers who Galileo stripped of their claims to authority. Certain people, still today, go on repeating the terse judgment of Gaston Bachelard: "Opinion is, in principle, always wrong. Opinion *thinks* badly; it doesn't *think*: it *translates* needs into knowledge."[2] That this judgment was emitted in a book entitled *The Formation of the Scientific Mind* assumes a profound logic. What is called the scientific mind only has a meaning in opposition to what would be nonscientific. Even if some think themselves clever in reversing the sense of the opposition, by attributing to people a subjective or emotional richness the absence of which would be characteristic of the cold, calculating, rational, scientific mind, as long as such an operation prevails, Science is recognized as having the power to extend its objective approach to everything that matters.

Not all scientists have adhered to the staging of "Science versus Opinion," which gives Science the role of defining the "real" questions, those that can be settled objectively, and of relating all the rest to subjectivity and its irrational attachments. But amongst those who know that this is only a matter of propaganda, some think that it was unfortunately necessary, otherwise the true value of the important work of scientists would not have been recognized. Moreover, this may be what Galileo thought. The contempt for people that the opposition between Science and Opinion propagates then takes softer forms – "they" cannot understand what we do, thus we have to offer them what will inspire in them the respect we are due – but the contempt is there nonetheless, simply in the fact that the price that has been paid, and continues to be paid, in order that the value of science being recognized might appear acceptable: it is acceptable to state neither the whole truth nor nothing but the truth, because people neither ask for nor merit it. People would lose confidence if one allowed them to know the extent to which a scientist is poorly prepared by his discipline for intervening in questions of a

2 Gaston Bachelard, *The Formation of the Scientific Mind,* trans. Mary MacAllester Jones (Manchester: Clinamen Press, 2001).

collective interest, and would throw themselves into the arms of charlatans, creationists, astrologers, and who knows what else.[3] A strange tolerance in relation to their colleagues, who endorse the role conferred on Science, characterizes most scientists, including those who know that the scientific mind or method that these colleagues are so proud of are the products of propaganda. A form of law of silence imposes itself from the moment that the colleagues in question seem to them to remain of good faith, even if this faith is blind.

I will come back to the knowledge economy, which is in the process of enslaving scientific practices but that will upset neither the scientific propaganda nor the authority of all-terrain "objective" proofs in the slightest. But from now on I want to underline the link between the sad passivity of the scientists who submit to this new management of public research, their inability to make politics out of what is happening to them, and this reference to Science, which, after having been so advantageous, is now strangling them. All they can do is whine about "the rising tide of irrationality" or that "they know not what they are doing", "killing the goose that lays the golden eggs." But this same passivity characterizes the so-called academic world in its entirety: it is in the process of being redefined by the very thing it allowed to be defined as "objectivity" when it was a matter of judging others – and I am including in this world those who protested against the "reign of objectivity." Whether it is a matter of the famous university rankings or the criteria of evaluation to which researchers and research centers are now subjected, they are produced by experts, who are also colleagues. And the facts that these experts record, which they identify as signs of excellence, may well be understood as blind, irrelevant or unjust – but they are no different to the ones that academics already accepted that other

3 Hence perhaps the excitement of many scientists faced with the creationist onslaught against Darwinian evolution: you see, here is the monster, and it is attacking us. So we are still a bulwark against obscurantism, like in the time of Galileo!

people would be subjected to, as objective, be it to endorse or to denounce the imperative of objectivity. The lack of resistance of academics against what is fabricating a new, operative definition of research cannot be dissociated from the way they accepted elsewhere generalized objectivity at its face value.

It is not a matter of academics complaining here, but of observing that the process of destroying the resources that might nourish an art of paying attention *continues unabated* under the cover of modernization, a process whose categorical imperative is the mobilization of all, with the door being shown to those who had till now benefited from relatively well-protected niches. Capitalism perhaps didn't demand quite so much, and it is here that the other protagonist, the State, shows itself. The resentful passivity of researchers comes in part from their feeling betrayed by this State, which they thought was in the service of the (well understood) general interest.

Not to complain, then, *but not to say* "it's only just" *either.* The intrusion of Gaia is opposed to this morality, which is in direct contact with the grand epic tale about the advent of Man: those who are unworthy, those who have been vulnerable to the temptations of the enemy, will only get what they deserve. I will repeat incessantly: we need researchers able to participate in the creation of the responses on which the possibility of a future that is not barbaric depends. It is an aspect of the GMO event that some started to manifest themselves; others will, undoubtedly. The way they are received matters.

Certainly I will not say that on the other hand we have no need of the State. I will rather say that faced with the intrusion of Gaia *the State must not be trusted.* It is a matter of abandoning the dream of a State that protects the interests of all, a bulwark against the excesses of capitalism, which is then only to be denounced for having failed in its mission. The question, then, is not one of knowing who (unduly) dominates the State and diverts it from the role it should play, which is the case when one talks about technocracy – whether "technology" here refers to engineering, to management, to science, or to law. Corruption obviously matters as well as conflicts of interest. But it seems more interesting to me – today especially, when the State's business is above all that of mobilizing for the economic war, without any credible reference to progress – to characterize what it is that the State does to the different technical practices that claim to serve the so-called general interest, what it does to those who busy themselves in its service. We know that their activity is most often characterized by the production of rules and norms (of quality, security, etc.) that are blind to locales and knowledges denigrated as traditional, and by the correlative elimination of what does not conform, is not standardized, what is recalcitrant in the face of objective evaluation. But to attribute all that to technical rationality is to go too quickly. As practitioners, technicians could be capable of many other things than subjecting everything that moves to categories that are indifferent to their consequences. The practices of a scientist, a technician, an engineer or a lawyer imply a particular art of attention, they allow (even demand, when they aren't enslaved) that they occasionally hesitate and learn. On the other hand, serving the State demands that there be no hesitation, it defines all hesitation as a threat to public order, as threatening demobilization.

But it is not for all that a matter of denouncing the State as an accomplice to, even as a direct emanation of, capitalism. However justified, denunciation fabricates a division between those who know and those who are duped by appearances. Worse, the knowledge that it produces has no other effect than to attribute even more power to capitalism. One could say, on the other hand, that between the modern State and its reasons, and capitalism, there is a chicken and egg logic. This entails not confusing the chicken and the egg – there is no symmetry between them – but affirms the impossibility of understanding the one without reference to the other, and vice versa, even if there is neither voluntary complicity, nor corruption, nor, moreover, friendship.

The first is forever complaining that "the State is too big." The other moans "we still need to impose some regulation." If it is a chicken and egg situation, it is because there is interreferencing of distinct logics of functioning, that of a machine said to be blind and heavy-handed, which defines what is entitled to be perceived and regulated, and that of an opportunist on the lookout, able to profit from everything that is not defined as perceptible.

Here again, I do not intend to define the logic of the State, but to try to characterize it, and this on the basis of what has happened. For nearly thirty years, our history has been that of the destruction of what was conquered through political and social struggle. Flexibility! Reducing red tape and state-imposed costs on employers! Everyone knows the quasi-consensual power that these demands by the bosses have succeeded in acquiring, the manner in which they have become order-words ensuring the weak adherence of the majority. But what has been so badly and so little defended was not what was conquered, but the *transposition into the categories proper to State management of what had been conquered.* I propose the term "whoever" to characterize this transposition. What has been conquered *for all* has been redefined by categories that are addressed to *whoever*, categories that produced amnesia and which are then vulnerable to the infernal alternatives concocted by capitalism.

Defeat, rather than victory (in this instance, the defeat of those who placed their trust in the State), allows these logics of functioning to be detected. In the era of social conquest, it was possible to attribute a progressive dynamic to the State, but when it turned and ran, it didn't betray anything. Its logic hasn't changed. Public order demands rules, and these rules demand a "logic of whoever," whoever designating all those to whom a rule or norm is to be applied *whatever the consequences of this application might be*. If there is an interreference between State and capitalist logics of functioning, between those who think themselves responsible for public order and those who clamor for a right to irresponsibility, the condition of free enterprise, it would pass via the hostility towards the art of paying attention to consequences that is common to them both, but for distinct reasons.

Of course, exceptions abound for every rule, and these exceptions are motivated by consequences to avoid. But they are always translated in terms of subcategories, or sub-subcategories, each time grouping together a class of "whosoevers," a class defined by the homogeneity of those that it includes from the point of view of the rule. And woe betide anyone who doesn't have the power to make their claim to be an exception heard. Woe betide, for example, the small farmers crushed by the administrative paperwork and regulations imposed in the name of consumer safety, if they haven't been able to make the case that only the large industrial farms can afford this cost. Woe betide too those who have been able to make themselves heard and have seen what they struggled for redefined in the terms of the State, and transformed into regulated functioning, blind to its consequences.

It goes without saying that big businesses, with their armies of lawyers and lobbyists, avoid the category of whoevers. Sometimes it happens that they do what "whoever" cannot do: obtain the adoption of rules that suit them, as was the case for Monsanto with the US administration with regard to the safety of GMOs, or get the State to act in their service directly, as is the case with the unilateral retaliation taken by the United States against countries judged lax with regard to respecting intellectual property rights. But more routinely, they are perfectly happy playing the game of whoever, that is to say, of benefiting from the legal fiction that makes them "moral persons," even being able to claim human rights. Except that they not only have the rights but also all the means they need to find the dodges that allow them to twist a rule or to make it work to their benefit.

Let it not be asked why the world of free enterprise continues to be opposed to the authoritarian, planner State. This is the alternative that subsists when the two crooks, State and (an Enterprise that can, consequently, be called "capitalist") reach an understanding to empty out the scene, to silence, or to ignore the voices of those who object, who demand that attention be paid to consequences that are unforeseen or haven't been taken into account or are intolerable. In short, of those who claim the capacity to intervene, to complicate matters, to meddle with that which – from the point of view as much of the State as of the Entrepreneur – doesn't concern, them. Especially not them.

If the question that matters today is that of a collective reappropriation of the capacity for and art of paying attention, the State, such as I have just characterized it, will not help. For the State, the springing up of groups meddling with what concerns them, who propose, who object, who demand to become actively involved in the formulation of questions, and learn how to become so, is in the first place always a "trouble to public order," which it is a matter of trying to ignore, and if that isn't possible, something about which amnesia will have to be produced. Public order, with its claims to being synonymous with the protection of a general interest that it is a matter of explaining to a population that is always suspected of wanting to give primacy to its selfish interests, reestablishes itself incessantly. We are swamped with consensual narratives, in which what has succeeded in counting is presented as normal, in which struggle is passed over in silence, in which those who have had to accept become those who "have (by themselves) recognized the necessity of...."

That is why attention must be paid to the contemporary appearance of "other narratives" that perhaps announce new modes of resistance, which refuse the forgetting of the capacity to think and act together that public order demands. I will devote myself here to narratives that make reference to "enclosures," that is to say, to the history of the expropriation of "commons."

[8]

Enclosures

"Enclosures" makes reference to a decisive moment in the social and economic history of England: the final eradication in the eighteenth century of customary rights that bore on the use of communal land, the "commons." These lands were "enclosed," that is to say, appropriated in an exclusive manner by their legal owners, and with tragic consequences, because use of the commons was essential to the life of peasant communities. A frightening number of people were stripped of all means of subsistence. "The Tragedy of the Commons" is, moreover, the title of a widely read essay published in 1968, but its author, Garrett Hardin, misappropriates the association between the destruction of the commons and tragedy. The tragedy is in fact supposed to be the overexploitation (postulated by Hardin) of the communal lands themselves, linked to the fact that each user pursued his self-interest without taking into account the fact that the outcome of this self-interest would be the impoverishment of everyone. This fable evidently met with great success, as it not only allowed the enclosures to be legitimated as "unfortunately necessary" but with them the ensemble of privatizations of what had been of the order of collective management: the interest of private property owners is also selfish but it pushes them to turn a profit on their capital, to improve their returns, and to increase productivity.

Another classic narrative – that of Marx – associates the expropriation of the commons with what he calls the "primitive accumulation of capital." The great mass of the poor, now stripped of any attachment, will be mercilessly exploited by the nascent industries, because there is no need to take into account the "reproduction of labor power": the poor can collapse on the job as there will always be others. In this sense, the enclosures prepare the capitalist appropriation of the labor of those who, deprived of their means of life, will be reduced to being nothing but their labor power. Marx, however, did not celebrate this expropriation in the manner in which he celebrates the destruction of the guilds and of the ensemble of what attaches humans to traditions and ways of life: like the elimination of an old order, an elimination that the future socialism will be indebted to capitalism for. Perhaps it is because of the pitiless brutality of the operation, or because what was destroyed was a form of primitive communism bringing resources and means into common use, but the fact remains that he saw in it a "theft" or the destruction of the "right of the poor" to ensure their subsistence.

If, today, the reference to enclosures matters, it is because the contemporary mode of extension of capitalism has given it all its actuality. The privatization of resources that are simply essential to survival, such as water, is the order of the day, as well as that of those institutions which, in our countries, had been considered as ensuring a human right, like education. Not that the management of water has not been a source of profit, and that capitalism hasn't largely profited from the production of well-trained and disciplined workers. What has changed is that henceforth it is a matter of direct appropriation, under the sign of privatization of what were public services.

And privatization doesn't stop there. The reference to enclosures is very directly activated by this knowledge economy to which I have already alluded, because what the latter promotes is nothing other than the disappearance of the line separating public and private research and the direct appropriation of what had, until now, benefited from a (relative) autonomy. The production of knowledge today is considered a stake that is too important to allow this minimal autonomy to researchers, who are now subjected to the imperative of establishing partnerships with industry, to defining acquiring a patent as the desirable success par excellence, and the creation of spin-offs as the glorious dream. All that with public money, which gets sucked up into multiple spin-offs that fail, whilst those that succeed will be purchased, without too much risk, together with their patents, by one or another consortium.

In short, the distribution between what the State lets capitalism do and what capitalism gets the State to do has changed. The State lets capitalism appropriate what was defined as forming part of the public domain, and capitalism gets the State to endorse the sacred task of having to hunt down those who infringe the now sacrosanct IP. Rights to such IP extend over practically everything, from the living thing to knowledge previously defined as freely accessible to all its users. Rights to which the WTO intends to subject the entire planet, in the name of the defense of innovation.

The contemporary reference to enclosures, to the appropriation of what was a common good, however, was not invented by union movements defending public services, or by researchers set to be run directly by their old industrial allies, with the blessing of the State. It was computer programmers, whose work was directly targeted by the patenting of their algorithms, that is to say, their very languages, who named what was threatening them thus, and created a response, the now celebrated GNU general public license. This was the point of departure for a movement for the collective creation of free software, which anyone can download and contribute to the proliferation of as competence and time allows. Let's not fool ourselves: it is not a matter of the angelic reign of disinterested cooperation. Other ways of making money were organized. But it is a matter of the invention of a mode of resistance to enclosure: everyone who has recourse to programs with a GNU license, or who modifies them, falls under the constraint of the exclusive non-appropriation of what they create.[1]

The resistance of programmers fits into the general category of struggle against exploitation with difficulty, because it is a matter of resisting the capitalism of the knowledge economy, and those who serve it rarely define themselves as exploited. Of course it is always possible to keep holding on to the theoretical compass, to maintain the heading that identifies capitalism and exploitation by speaking of a form of "false consciousness" – they do not know that they are exploited, but we do. Sticking to the heading here, however, amounts to denying the originality and relative efficacy of what programmers who resist have succeeded in doing. If they had joined the struggle of the exploited masses, IP rights would reign undivided today over the domain of software.

How is this type of resistance, which has transformed the reference to the commons as a stake in a struggle, to be recounted? I will distinguish two types of narrative here, in a manner that is a little caricatured, certainly, but it is the divergence that it is a matter of making sensible here, not the positions themselves.

---

1 That I am referring to the free software movement here doesn't signify that they are "good," whereas the software "pirates" and "crackers" of protected software who distribute pirate copies that avoid protection are without interest. One might say that at the level of effects – their power to harm the property rights and ensure the free access to programs – the pirates are more effective. But there is no need to choose here – many in any case belong to both milieus. Nor is there any need to oppose them, like one might oppose reformism and radicalism. Both movements are interesting, neither is exemplary (if many creators of free software get on well with profit, gratuitous piracy, like every war machine, communicates with a problem of capture: many such pirates are taken on as experts and become hunters).

The first narrative stages a renewal of the Marxist conceptual theater, which preserves the epic genre (characterizing it in this manner is a way of announcing that for me it is a matter of distancing myself from it).[2] Capitalism today supposedly has to be qualified as "cognitive" – it aims less at the exploitation of labor power than at the appropriation of what must be recognized as the common good of humanity – knowledge. And not no matter what knowledge – it is the workers of the immaterial, those who manipulate abstract knowledges in cooperation with one another, who have become the real source for the production of wealth. From now on, this "proletariat of the immaterial," as Toni Negri says, is what capitalism is going to depend on but is what it will (perhaps) not be able to enslave. Because the specificity of immaterial knowledges, ideas, algorithms, codes, etc., is that their use value is immediately social, as is language, already, which only exists by and for sharing and exchange. The new enclosures would thus translate this new epoch, in which for capitalism it is a matter of preventing a social dynamic on which it now depends and that escapes it. And reciprocally the mobile and autonomous immaterial proletariat could well succeed at doing what the old peasant communities, attached as they were to their communal fields and their concrete knowledges, could not. The revolt of the programmers, the manner in which they have succeeded in constructing cooperative networks, which affirm the immediately social value of the immaterial – because every user is now, thanks to them, free to betray Bill Gates and to download the programs he or she needs – would thus be an exemplary annunciation.

It is thus still a matter of an epic for humanity, a humanity to which capitalism has, in spite of itself, revealed its true vocation.

2 The proposition from which I am distancing myself here is that of Toni Negri and Michael Hardt, staging what they called the "multitude." This multitude, which is fundamentally anonymous, nomadic, and expert, becomes the new antagonistic force capable of threatening capitalism. The latter, become cognitive, has a vital need of the multitude, which is, on the other hand, capable of escaping its grasp, because it is not identified with industrial modes of production.

To the extent that cognitive capitalism exploits a language that allows the communication of everyone with everyone else, a knowledge which, produced by each benefits everyone, would make exist, here and now, what is common to humans, a common that is fundamentally anonymous, without quality or property. Without wanting to, capitalism would thus contribute to the possibility of a humanity reconciled with itself, a mobile creative multitude, emancipated from the attachments that brought groups into conflict. And as there can only be one revolutionary epic, the working class is chased from the role that Marx had conferred on it, indeed, is even defined in terms that retrospectively disqualify it from playing that role. It was supposed to have nothing to lose but its chains, but those who have lost their chains already exist – or have already acquired a conceptual existence, at least. And the old working class itself, whose work was material, is now characterized as being too attached to the tools of production to be able to satisfy the concept, to be a bearer of the "common" of humans.

From the conceptual point of view, the fact that in the name of competition workers are exploited today with a rare intensity, without even talking about the *sweatshops* reserved for poor countries, or about the appearance in our countries of poor workers who aren't capable of making ends meet on their salaries, doesn't count for much. But above all, as in every theater of concepts, we are functioning here in the long, even the indefinite, term. Mathematicians might talk about a theorem of existence: what is conceptualized demonstrates the existence of a positive answer to the question "is there a candidate worthy of the role?" but doesn't indicate the manner in which the candidate will become capable of fulfilling this role. It is precisely this kind of research, for a conceptual guarantee, that Gaia interrupts, and does so in the most materialist mode there can be. The response to her intrusion will not admit, cannot admit, any guarantee, because Gaia is deaf to our ideas.

Let us take up again the direct appropriation that programmers have been able to resist, these enclosures that were to suppress their own manner of working and cooperating. Might they not remind us of another dimension of capitalism, not one that is concurrent with exploitation but required by it and, as such, propagated wherever new resources to be exploited can be envisaged? According to the second narrative that I am proposing, what was destroyed with the commons was not just the means of living for poor peasants, but also a concrete collective intelligence, attached to this common on which they all depended. From this point of view, it is this kind of destruction that programmers have been able to resist. They would no longer be the figure of annunciation, represented by the immaterial nomadic proletariat, incarnating the common social character of immaterial production. The "common" that they were able to defend *was theirs*, it was what made them think, imagine, and cooperate. That this common may have been immaterial doesn't make much difference. It is always a matter of a concrete, situated, collective intelligence, in a clinch with constraints that are as critical as material constraints. It is the collective brought together by the challenge of these constraints, rather different from the indefinite ensemble of those who, like me, use or download what has been produced, that have been able to defend against what had endeavored to divide them. In other words, the programmers resisted what was endeavoring to separate them from what was common *to them*, not the appropriation of the common good of humanity. It was as "commoners" that they defined what made them programmers, not as nomads of the immaterial.

The divergence between the two narratives thus bears on the question of community. From the point of view of the first, there isn't any great difference between the creators and the end users of software, like me – we all have in common this abstract language of a new type, belonging to no one, free of the attachments that divide, that oppose, that make for contradictions.

From the point of view of the second, cognitive capitalism doesn't appropriate the inappropriable, but destroys (continues to destroy) what is required by the very existence of a *community.* The "common" here cannot be reduced to a good or a resource and it doesn't in the least have the traits of a sort of human universal, the (conceptual) guarantor of something beyond oppositions. It is what unites "commoners," I utilize software as an end user, but those who resisted enclosure by IP rights did not defend the free use of a resource but the very practices that made them a community, that caused them to think, imagine, and create in a mode in which what one does matters to the others, and is a resource for the others. And it is as such, because the knowledge economy was attacking what made them a community, and not as the precursors of a multitude freed of its attachments, that they laid claim to the precedent of the enclosures.

[9]

Common Causes

To use the word "commoner" to talk about programmers who have been able to resist is thus to situate them in the lineage of peasants who, in the past, struggled against the confiscation of their commons in a mode that no longer defines these peasants as poor but as communities. And it is also to associate this resistance with the recent political creation called "user movements." This is what takers of illegal drugs called themselves, in a movement in which they created an expert knowledge with regard to this practice, and called for this expertise to be recognized as such by the "experts." There has been something similar for patient associations faced with doctors and pharmaceutical enterprises. But the term has also been used to talk about those who unite around a "common," a river or a forest, with the ambition of thwarting the sinister diagnosis of the "tragedy of the commons" and of succeeding in learning from one another not to define it as a means for their own ends but as that around which users must learn to articulate themselves. In each of these cases, and there are many others, the success of the movement derives from those who were initially defined as utilizing something, seizing hold of questions that they weren't supposed to meddle with, and *conferring* on the "common," which was often defined in terms of rival utilizations, *the power* to gather them, to cause them to think, that is to say, to resist this definition, and produce propositions that it would otherwise have rendered unthinkable. In brief to learn again the art of paying attention.

One must not go too quickly, however, because the rapprochement of programmers and commoners quickly encounters difficulties that it would be dangerous to ignore. Whether they have resisted or not, programmers know that, like scientists or lawyers, they are bearers of a recognized knowledge, which makes them what I call practitioners. On the other hand, in the same movement, unrepentant drug users and members of associations such as Act Up, for example, have created a collective "profane" knowledge and struggled for the recognition of this knowledge by practitioners and acknowledged experts. That they may be able to succeed in transforming the latter, forcing them to pay attention to the dimensions of a situation that haven't been taken into account, certainly matters, but isn't confusing these two types of protagonists under the same term to introduce an ambiguity regarding its signification?

I haven't stopped emphasizing that the question I am posing is not "what is to be done in the face of the intrusion of Gaia?" – a question whose answer belongs to the multifold process of its creation – but "what does trying to respond to the intrusion of Gaia in a mode that isn't barbaric call for?" Such a response will need the contributions of scientists, technicians, and lawyers, but not those of people who work under the yoke of the knowledge economy, nor those of people who define themselves, one way or another, by a contempt for "people." That is why I consider that the type of ambiguity I have just arrived at, or more precisely, that the resistance of programmers as much as the creation of user movements, have allowed me to arrive at, is precious. The fact that I am tentatively using the same term "commoners" for practitioners who defend what causes them think and imagine, and for the heterogeneous group of those who learn to be caused to think by what they refuse to be the end users of, creates an ambiguity that doesn't have to be removed but much rather made explicit. To remove it would be to look for a ready-made solution, and there is no such solution when the question is "making common." This question must rather be a dimension of situations that, around a common concern, gathers representatives of user movements, practitioners, and experts, a dimension that belongs to the situation and cannot be thought independent of it.

I am alluding here to a difficulty that is well-known in user movements that have gained the right to intervene in technical discussions from which they were excluded. This moment of relative success, the moment that one moves from a position of contestation to a position of having a stake, is also the moment of greatest danger. In order to learn how to address themselves to practitioners and experts, those who participate in such discussions must learn how to get to know them, to get the measure of their knowledge, and this necessity is often the source of great tension. The users' engagement around a common cause is put to the test by a divergence that can be actualized in personal conflict. Suspicions bearing on the ambitions of some – "you talk like them, you've become one of them" – will be met with reproach regarding others' lack of investment – "is it my fault if I'm the only one to make any effort? You only had to...." Making the ambiguity explicit is not to resolve the difficulty. There is no general solution here, the only generality is *the necessity of foreseeing that there will be tension*, that is to say, in particular, of nourishing the common engagement with knowledges, narratives and experiences which, when the time comes, will allow the trap not to be fallen into.

We will not, however, oppose practitioners, who would be people with a genuine craft, and users, who would be amateurs who wish to assert their objections and suggestions but would be divided when it is a question of participating fully in the construction of the problem. The question of divergent engagements is equally posed on the practitioners' side. They too can be divided, depending on whether they behave as professionals or are actually able to understand their specialized knowledge as contributing to a common concern, not defining it. In the first case, users will be dealing with interlocutors who will certainly agree to envisage the manner in which objections and suggestions can be understood, but who will already know how to pose the problem – users will therefore be heard as intervening at the level of the solution, not in the formulation of the problem, and those who enter into this game really will be in danger of being separated from the others, caught in insurmountable conflicts of loyalty. In the second case, it is not impossible that they might, with and in the same way as practitioners, contribute to the construction of the problem, the concerning situation now being defined in terms of the *heterogeneous* knowledges, requirements, and manners of paying attention that its unfolding demands.

The intervention of users thus activates a contrast that matters when it is a question of the contribution of those I call practitioners to the response when facing the intrusion of Gaia. And this contrast henceforth constitutes a political stake in the same way as does the distinction between users who participate in a movement and end users who defend their interest. Whether it is a matter of the end users or of those one calls true professionals, we are dealing with those whom Entrepreneurs and those who are responsible for public order can count on not to hesitate. But the question of practitioners has an extra dimension. One can *become* part of a user movement but one must be *trained* in a practice. This doesn't signify any kind of hierarchy but translates a belonging, the fact that the knowledge of a practitioner, her capacity to participate in the construction of a problem, refers to the community to which she belongs. The extra, political dimension is that a future in which the very notion of a practice would be destroyed, in which the sciences *would no longer produce anything other than professionals*, incapable as such of dealing with what the encounter with users demands, is easily imagined.

When, some years ago, I decided to question the sciences on the basis of the persona of the practitioner,[1] this was, in the first place, so as to resist the direct link so often established between Science and a neutral, universal rationality, but also so as to announce the inevitable conflict that would arise once demystifying critical studies began to show that scientists do not obey these famous standards of rationality. Demystification always risks throwing the baby out with the bathwater, here denying that some sciences (not all) actually bring new reliable knowledge about the world, and populate reality with new beings and agencies. To speak about scientific practices was meant both to characterize their own specific force, irreducible to general social relations, and to unlink this force from any claim to a rationality that would be lacking amongst nonscientists. That is why I have tried to characterize scientific practitioners (in contrast to those who serve Science) as gathered together by a "common," that is to say, by a cause: they are *engaged by a type of achievement* proper to each field *the eventuality of which obliges those who belong to this field*, forces them to think, to act, to invent, to object, that is to say, to work together, depending on one another.

Today, it has to be noted that scientists have not, in the manner of programmers, invented a manner of resisting the enclosures that are their lot too in the knowledge economy. That this is paid for by a loss of reliability can already be sensed, with the multiple cases of conflicts of interest – when one discovers that a scientist who presents himself as an expert on a question benefits from subsidies from an industry interested in this question. But even when there is no direct conflict, the situation of dependency is enough to destroy reliability because it dissolves the obligation

1 Practices and practitioners are introduced in my *Cosmopolitics* vol. 1 and 2 (Minneapolis: University of Minnesota Press, 2010–11) and envisaged from the point of view of an ecology of practices in *La Vierge et le neutrino*.

 to work together. One can *"succeed" differently, with completely different means*. Soon the baby will be thrown out with the bath-water and demystification will be redundant. We will be dealing with "true professionals," who do not hesitate and who do not fear the objections of their colleagues. Because when everyone is dependent, when everyone is linked by partnerships to industry, no one will want to "spit in the soup," to carry out research that might weaken the legitimacy of their, and everybody else's, participation in the industrial redefinition of the world. There is no need for trickery, it is enough to avoid working on questions that are challenging and to focus on those for which grants and public support are abundant. Even if, as is the case with nano-technology, it means shifting from a knowledge economy to an economy of promises. In this instance scientists promise the moon on a stick, a new industrial revolution, a new age of Man, no longer taming matter but atoms, assembled at whim one at a time. They do not fear objections on the part of their colleagues any longer, and industry and public powers join the somewhat obscene merry-go-round, where no one knows who believes, who is dupe, or who manipulates who any longer....

What is in the process of happening with the knowledge economy demonstrates well the associating that I am attempting to make between enclosures and the destruction of practices generically taken as production of collective intelligence. What will be destroyed is not just the communities of practitioners, united by a cause that leads them to think, imagine, and object. In effect, what distinguishes practitioners from professionals is also the capacity to perceive the difference between situations and ques-tion the definition of what matters to them as a community, what causes them to gather, and to others for which their knowledge or expertise can be useful, even necessary, but will never allow them to define the "right manner of formulating the problem." Certainly, and it's the least one can say, such a capacity hasn't really been cultivated by scientific communities and the modes of training they developed. But with the triumph of professionals, this capacity will be eradicated. Another potential resource will have been destroyed, which matters in a crucial manner if it is a question of the gathering together of heterogeneous knowledges, requirements, and concerns around a situation that none can appropriate.

From the point of view that she poses the question of our capacity to create responses that aren't barbaric, the intrusion of Gaia gives a formidable significance to the destruction of common causes that I have associated with the enclosures of yesterday and today. And she gives a crucial sense to the double distinction that I have proposed between users, or commoners, and end users, between practitioners and professionals. We urgently need to learn how to resist the nasty little song that sweetly whispers "that's what people are like (selfish, subjected to their habits of thought, etc.)", this little song whose theme is what intellectuals call voluntary servitude – which is always that of others, of course. No, the transformation of users into (selfish) consumers, or practitioners into (submissive) professionals doesn't testify to people always being inclined to follow the easiest path. It testifies to the destruction of that which gathers together and causes people to think. But to adopt this point of view is equally to take note that the response to intrusion will not be one that a humanity which is finally reconciled, reunited under the sign of a general goodwill, would become able to give, but depends on the *repopulating* of a world devastated today by the confiscation or the destruction of collective, and always situated, capacities to think, imagine, and create.

From this point of view, what the GMO event was able to yield matters: to make our guardians stammer, to make the evidence on which they count to lead their flock towards a future that they themselves are incapable of conceiving lose its hold. The question of knowing how they might do otherwise, without anesthetizing order-words, is a different story, which is not ours. What we now know is that our hypothetical future, the stories through which a response to Gaia could be created, doesn't pass via the taking of the Winter Palace or the Bastille. It is not a matter of a refusal of a moral type, refusing to take power so as to keep one's own hands clean. The question is rather technical: "taking power" presupposes that a government has power, that it can betray the role that capitalism makes it play. How *to reclaim power* is doubtless a better question, but the response then passes via a dynamic of engagements that produce possibilities, a dynamic that breaks the feeling of collective impotence without toppling over into the formidable "together anything becomes possible!"

Breaking the feeling of impotence in effect has nothing to do with what is, rather, the correlate of impotence, the feeling of omnipotence, the cult of hidden powers that ask only to be liberated, the abstract dream of the day when, at last, "the people will be in the street." If it isn't only a question of the reappropriating of the wealth produced through work, the people who may well invade the street should come there with concrete experience of what is demanded by reclaiming what has been destroyed, reappropriating the capacity to fabricate one's own questions, and not responding to the trick questions that are imposed on us. One never fabricates in general and one is never capable in general.

The people in the street is an image that I do not want to give up, however, because it is an image of emancipation that can be delinked from the grand, epic prospect. After all, before our cities were reconfigured according to the imperatives of frictionless circulation, purified of threats to the public order that crowds and mixing together can always constitute, the people were in the street... But to prevent this image from becoming a poison, an abstract dream, perhaps it is worth transforming the image of what a street is. For the grand boulevards that lead to the places of power, a labyrinth of interconnected streets could be substituted, that is to say, a multiplicity of gatherings around what forces thinking and imagining together, around common causes, none of which has the power to determine the others, but each one of which requires that the others also receive the power of causing to think and imagine those that they gather together. Because if a cause is isolated, it always risks being dismembered according to the terms of different preexisting interests. And it also risks provoking a closing up of the collective, the collective then defining its milieu in terms of its own requirements, not as that with which links must be created. Which is what has happened to scientific communities. In short, a cause that receives the power to gather together is, par excellence, that which demands not to be defined as good, or innocent, or legitimate, but to be treated with the lucidity that all creation demands.

[10]

It Could Be Dangerous!

Some eyebrows might well be raised at the prospect that I have just opened up. After all, my example bearing on the sciences cuts both ways. Well before entering the stage of the knowledge economy, did not scientists conclude privileged alliances with industry, the State, and the army? And have they not contributed, since the nineteenth century at least, to the type of development that has for us merited the intrusion of Gaia? Have they not played on their authority so that the undesirable or threatening consequences of this development would not be taken into account, in the name of future progress that would repair the damage, or even more simply, as the price of progress? In other words, do they not offer the example of what happens when one obeys not the common interest, but one's own interests, whether or not they are those of a practice?

Certainly one retort could be that scientists have, as far as what didn't concern their own practices goes, shared the great trust of the majority (a majority amongst those who felt themselves qualified to speak in the name of humanity...) as to the irresistible drive of Promethean Man, he who breaks limits and ignores prohibitions. But the objection goes further. Because with the example of scientists, it is the manner in which I proposed to associate commons with a capacity for resistance, for the reclaiming of capacities to think and act together, which raises eyebrows. Doesn't the vulnerability of scientists to the grand narrative that they were the heroes of, which made of them the collective brains of humanity, demonstrate that I am placing my trust in the collective intelligence that would characterize those I am calling practitioners or users really too easily? Certainly we live in a veritable cemetery for destroyed practices and collective knowledges, but is it for all that necessary to entertain an idyllic vision of these commoners united by and around a common? Is it not necessary to fear corporatist reflexes? In short, have I not fallen for a typical illusion, one incessantly denounced by Marxists, namely the trust in a spontaneous capacity for resistance that needs neither theory nor compass, and in which it would be necessary to trust?

The objection matters, and it is now necessary for me to underline that the characterization that I am trying to link to the theme of the enclosures – that of a capitalism that isn't just an affair of exploitation but which requires, and doesn't stop propagating, an operation of destruction – *does not signify* that those whom I have called practitioners or those who call themselves users, offer as such any guarantee whatsoever of reliability.

In fact, those who object will be able to line up the most disastrous of examples. They could evoke the trap laid for workers when they have been associated with quality circles from long ago (already), in which it was a matter of thinking together about how everyone could contribute to the common cause that the good of the enterprise constituted. They will also be able to evoke the reasons why the unions who represent public service workers scorn any alliance with users, knowing that the latter can very easily propose reforms that would upset relations of force established with difficulty, to the detriment of workers. In another register one may think of patient associations who have become the best allies of the pharmaceutical industry, calling for a distance from the norm (hyperactivity, for example), for which this industry has precisely provided medication, to be recognized as a real illness. But above all one can see looming the general question of the unraveling of politics to the benefit of governance by stakeholders, those who have an interest (a share) in a situation. Despite my fine assurances, can one not hear the murmuring of the great refrain of stakeholders: "let the others collapse, let all the rules that aim to avoid deepening inequality disappear, we demand to be able to play all the winning cards we possess in 'free and undistorted' competition," the credo of the European Union (EU).

In short, to evoke the commoners, practitioners or users, those whom a common cause unites, those who have to give to what they all depend on the power to cause them to think together, albeit in different modes, *is not, in effect, without danger.* And the first danger is to evoke them as the source of unprecedented alternatives, enabling resistance to the capitalist takeover of the future. One could even see here a new version of the fascistic opposition between the "real country," perfectly able to take its future in hand, and the clique of those who confiscate its power to act and determine itself. Any naïvety in the matter could be disastrous.

Nevertheless, one must equally resist the "and so..." that follows all too rapidly the disqualification of those who announce the good news, the discovery of the human capacity to self-organize, the hidden resource that will resolve everything. Because this "and so" brings discredit to the experimental efforts that always laboriously, sometimes messily, seek effectively to produce this collective intelligence. Both those who announce the good news, as well as the skeptics and the anxious, who make an argument from the dangerous drifting to which such efforts are vulnerable, contribute to their weakening, like an unhealthy environment that infects those who try to live in it.

I would maintain that the question of what the commoners need – have a crucial need of – is a particular version of the art of paying attention. It is a matter of the art of what the Greeks called the "pharmakon," which can be translated as "drug." What characterizes the pharmakon is at the same time both its efficacy and its absence of identity. Depending on dose and use, it can be both a poison and a remedy. The type of attention that their milieu can lend to user movements is a pharmakon. It is capable of both nourishing and poisoning them. And the same "pharmacological" uncertainty prevails with regards to what these movements themselves can produce. That they might be dangerous thus goes without saying – every pharmakon can be dangerous. What it is a matter of putting into suspension, through referring to the instability of the pharmakon – remedy or poison – is the way this danger functions *as an objection.*

When one of our guardians cries – and this is the cry by means of which we recognize that he effectively thinks of himself as responsible for us – "but that could be dangerous!" he inherits with this "but" a history in which the instability of the pharmakon has been used again and again to condemn it. A history in which what has been privileged again and again is what presents, or seems to present, the guarantees of a stable identity, which allows the question of the appropriate attention, the learning of doses and the manner of preparation, to be done away with. A history in which the question of efficacy has been incessantly enslaved, reduced to that of the causes supposed to explain their effects.

The hatred of the pharmakon goes back a long way. If one wishes, one can trace it all the way back to Plato, who defined philosophy by the requirement of such stability against its sophist rivals, who were capable of the better and the worse. Or Christian monotheism, inventing an intrinsically good God. Or the question of the power of judgment, which needs to be able to abstract out from circumstances. Or even the passion for recognizing genuine claims from amongst imposters, a passion that nourishes a certain thirst for the truth. In any case, our history is saturated with multiple versions of the same obsession, that of doing away with the pharmakon and retaining only that which offers the guarantee of escaping from its detestable ambiguity. But it may well be that privileging what would offer such a guarantee, as it lures us into not paying attention, provoking the imprudence of an unthinking use, stabilizes the efficacy as a poison of what is defined as a remedy.

Let us come back, from this point of view, to the contrast between the response that programmers were able to give to the operation of enclosure that threatened them, and the passive resentment of most of those scientists who have not already embraced the imperative of the knowledge economy. It is a contrast that is all the more intriguing because it was the cooperative character of scientific research that served as a reference for the programmers. Why have programmers not only succeeded in defending their capacity to cooperate, but also to think and to invent links with end users who, like me, now count on the possibility of the free download of software that meets their needs? Why have scientists preferred to link themselves with States and entrepreneurs, and why have they defined the rest in terms of a lack (a lack of knowledge, a lack of rationality) or of a fear (of change, of challenging the unknown), in such a way that at the moment their allies started to enslave them, they found themselves incapable of imagining a possibility of resistance?

To think in pharmacological terms here is to pose the question, not of the identity of the sciences, *but of the differences in milieu* of these two practices, milieus that are not only external but that include the manner in which practitioners evaluate their relations with them. From this point of view, the event that the "birth of modern science" constituted is significant. Today one still finds authors, who are nevertheless interesting, who keep on repeating the stupid error that the explosive development of Europe, in contrast, most notably, with China, was due to the discovery of the power of scientific rationality, enabling it to identify the laws that nature obeys. The success of the propaganda operation initiated by Galileo, and which still infects the imagination of scientists as much as nonscientists, may well derive from the propaganda in question having almost no need of propagandists. The practical novelty effectively associated with experimental proof may well have found a milieu already prepared to echo it in this way. As rare and restricted in scope as the experimental facts, able to testify to the manner in which they must be interpreted, are, this capacity may well have reactivated the old hatred of the pharmakon, of unstable opinion, of undecidable interpretations. A relationship with the world that was at last rational had been created!

What the propaganda fed upon then would be less the novelty of the experimental success and more the satisfaction of a much older requirement, the requirement that a truth imposes itself, that is to say, is able to manifest its difference from its rivals. Consequently it is no surprise that the "it has not been proven" came to be associated so easily with "it is thus not worth counting," indeed with the suspicion of irrationality coming to weigh upon those who took an interest in what has not been proven.

By contrast, one could say that from the outset the practice of programmers was placed under the sign of knowing that what they produced could be a remedy or a poison – notably, under the sign of a possible future in which Big Brother reigned. And the correlate of this contrast is the singularity of the history of practical innovations within information technology. It is a matter of a rare case in which the technical, cultural, social, and political stakes are intimately linked. A case that is all the more remarkable for the anchoring of this history in a military development. It is not, in effect, a matter of forgetting that information technology is entwined with war, or that today, more than ever, it is an instrument of control, repression, and exploitation. But that that is *not all it is* is something that is perhaps owed to this particularity of the practitioners, who never thought that their technique was innocent, who never made the choice of its good or bad use the responsibility of the politician – see the celebrated argument ritually used by scientists: is it the fault of the inventor of the axe if it has been used for killing?

The pharmacological approach doesn't permit the question of whose fault it is to be the crucial one, the distribution of guilt and innocence to be an aim in itself – programmers who have been able to resist are not better than scientists who haven't. But it proposes "thinking by the middle/milieu." And the case of the scientists shows that a milieu obsessed by a stable distinction being established between remedy and poison is a milieu that empoisons, which even destroys. How many efforts have been disqualified because they couldn't offer guarantees that no one should be capable of offering?! How many false, illusory guarantees have been offered and accepted at face value?! How many brutal judgments have been passed with regard to that which, being fragile and precarious, asked to be nourished and protected?!

In any case, the time of guarantees is over – that is the first meaning to confer on the intrusion of Gaia. This does not signify that anything goes, a resigned sigh or the horrified cry that express again and again the search for a value endowed with the power to denounce its rivals, who would be nothing but frauds. It does signify that what is valuable must in the first place be defined as vulnerable. By definition the dynamics of the creation of knowledges, of struggles, and of experiments that will respond to the intrusion – each insufficient by itself but important through its possible connections and repercussions – will be vulnerable.

A response cannot be reduced to the simple expression of a conviction. It is fabricated. It succeeds or fails. No manner of responding has to proclaim a legitimacy that transcends circumstance, that demands recognition on the part of all, that dreams or requires that all accept it as determining. But nor can any be condemned because it might be vulnerable to drifting dangerously. What the art of the pharmakon proposes to those who posit the diagnosis "it could be dangerous" is, by contrast, to recognize that the objection engages them, makes them an integral part of the process of fabrication. If they want to ignore that they are an integral part, they will still be so, but as judges who will contribute to a hostile or ironic milieu. On the other hand, they can also be so as allies, with questions like "how can we contribute to avoiding this danger?," "how are we to cooperate against what will be employed to confirm our diagnosis?" and "how can we participate in the creation of a milieu that will help what is venturing to exist?"

There is but one certainty: that the process of creation of possibility must be very careful of the utopian mode, which appeals to the surpassing of conflicts and proposes a remedy the interest of which must be respected by everyone. And there is but one generality that holds: that *every creation must incorporate the knowledge that it is not venturing into a friendly world but into an unhealthy milieu*, that it will have to deal with protagonists – the State, capitalism, professionals, etc. – who will profit from any weakness and who will activate all the processes likely to empoison ("recuperate") it. For example, by recognizing users in a mode that transforms them into stakeholders, by setting up situations that divide those who seek to cooperate, by demanding inappropriate guarantees, or by fabricating infernal alternatives that dismember that which was seeking to create its own position.

As I have already emphasized, the intrusion of Gaia upsets the order of temporalities. The pharmacological art is required because the time of struggle cannot postpone the time of creation. It cannot delay until "after," when there is no longer any danger, the time when humans will be able to unfold their creative capacities – life, thought, joy – and conjugate their efforts for the benefits of all. But it is also required because those who are seeking to create cannot do so innocently, by accusing those who struggle of wanting to take power whereas they would have known the need to turn their back on such an ambition. The times of struggle and of creation must learn to work together without confusion, through relaying, prolonging and reciprocal apprenticeship to the art of paying attention, on pain of mutual poisoning and of leaving the field free for the coming barbarism.

[11]

A Threat of Regression?

Conjugating struggle and creation without confusion sounds very good. Too good perhaps. Writing this essay, my aim is not to offer propositions that demand adherence, but to seek to put into words, and perhaps into thought, the manner in which what I have named "the intrusion of Gaia" puts our propositions to the test. It is thus a matter of stimulating something completely different to adherence – it is necessary instead that it grates, that it resists, that it protests. That in any case is why there is something deliberately provocative in my choosing to name Gaia, to designate her as an unprecedented, or forgotten, form of transcendence. It is a matter of a provocation that doesn't seek to scandalize, hence my precautions and explanations, but which nevertheless means to stimulate a minimum of perplexity or discomfort. Thus some may ask why – if what I have called Gaia asks nothing of us, if it is not a matter of a cult or of conversion – give it this name? Why employ the term "transcendence"?

What finds its expression in this perplexity or discomfort can be called a fear of regression, and this fear is long-standing, even amongst those who no longer believe in progress: there are things in our heritage that must not be renounced. But it is here that one must pay attention. Is the fear to which the refusal to renounce responds the fear of being *oneself* tempted to renounce? Or is it the fear that *others* may be drawn into renouncing? That's an entirely different matter.

The distinction that I have just brought about implies a properly pharmacological test. To fear on behalf of others is to maintain the position of the "brains" of humanity, thinking for and in the name of those who are supposedly vulnerable to temptations from which they must be protected. I will come back later to this fear, which I consider to be a poison that it is a matter of learning to recognize and resist. But I want first to address myself to the fear of regression in a mode that is appropriate to the painful perplexity of those who would wonder if, despite my assurances, I am not in the process of inciting a betrayal of that for which fidelity must be maintained.

It is impossible for me to speak for others about what they want to be faithful to. I will therefore speak for myself, refusing to turn my back on that important moment in European history that is called the Enlightenment, that moment when a taste for thinking and for the imagination as exercises in insubordination became widespread, in which a link of a new type between life and possibilities was forged. I do not wish to renounce that Enlightenment, and I want nothing to do with those who deny its happening, in the name of its limits and ambiguities.

I take myself as a daughter of the Enlightenment, then. But it belongs to those who identify themselves as inheriting such an event to ask the question of *how* to inherit it, that is to say too, how to avoid being its rentier, the representative of an established privilege that it could never be a matter of going back on, except by regression. Or else how to inherit the insolent laughter, the audacity of a Diderot, against the scientific mind that also claims to be an inheritor of the Enlightenment, but in the name of which the insolent are silenced. And above all, how is one to *treat*, in the pharmacological sense of the term, that which, since the Enlightenment, has been honored as the remedy par excellence for the erring of humanity, the *critical thinking*.

Let me be clearly understood: the question is not in the slightest one of contesting the utility, and even the necessity, of what is always an ingredient of thinking anyway, but rather the identification of critique as a remedy, that is to say also *its transformation into an end in itself* – an end in itself that would singularize we inheritors of the Enlightenment, amongst all other peoples. It is this transformation that generated the great epic genre in which Man becomes adult, takes his own destiny in hand and shakes off the yoke of illusory transcendences. The adventure of the Enlightenment then became a mission: at one and the same time a merciless combating of the monsters that incite us to regression, and a mandate to have to bring light to anywhere obscurity is said to reign.

Here I want to try to make those who feel themselves to be engaged in this combat hesitate. I will first emphasize that such combat is not associated with too many risks in our countries, where we call ourselves modern and where it is now extremely rare for the critical hero to provoke a ferocious raising of the defenses on the part of those whose illusions he aims to destroy. Today the exercise of critique has become a pastime for academics, who are not widely known for their courage, and a well-worn path for beginners whose doctoral dissertations kick over the statues of the beliefs supposed to dominate us again and again, to indifference and general tiredness. In some cases those who should be called the rentiers of the Enlightenment strike a heroic pose because they have provoked anger or hate. Our right to blasphemy is in danger, we hear. The question is not one of defending hateful reactions, but of underlining the indignity of this supposedly privileged right: to blaspheme has never meant insulting the belief of others who are distant, but those who are near, sometimes even our own beliefs. That is to say, it means running the risk of rejection, exclusion, or denunciation by one's own kin.

It would be easy to say that this risk of rejection is one that I run on the part of those who would accuse me of favoring regression, or of demobilization in a world in which the enemies of the Enlightenment are waking up again. But this kind of retort is inappropriate, to the extent that I am addressing those who I imagine hesitating, asking themselves if giving up the power of critique, its capacity to destroy illusions, is not about giving up the only defense that we have in a world in which illusionists proliferate.

On the other hand it is possible to share with those who hesitate the question that this epoch poses, in which it is the very possibility of progress that is being shelved in the store of lost illusions. Will not the barbarism that could well define the future be that which designates as illusions the finally dispersed causes that made live, hope and struggle those we want to inherit from? Is that not what we are already having much more than a foretaste of today, when the hold of capitalism, nonetheless rid of its pretences at bringing progress, is stronger than ever? In *Capitalist Sorcery* we wrote that "if capitalism were to be put in danger by denunciation, it would have collapsed long ago."[1] To which I would add today that barbarism doesn't fear critique. Rather, it nourishes itself on the destruction of that which appears retroactively as a dream, utopia, or illusion, as that of which reality imposes the renunciation. It triumphs when the memory of what has been destroyed is lost or makes people cackle or sigh.

The argument, however, would be insufficient if it was to be understood in the mode of tolerance, the necessity of suspending the critical weapon in order to allow all sorts of archaic or New Age beliefs to nourish a resistance to this reality. That is why it is necessary to go a bit further, and call into question the image of illusion that the heritage of the Enlightenment has been referred

1 Pignarre and Stengers, *Capitalist Sorcery*, 11.

to: illusion would be what veils the light, what separates us from truth. What that truth is depends on the spokesperson, but the point of convergence is the imperious necessity of dissipating the fog, of unveiling, demystifying, and not being duped. Now what is striking, in our modern countries at least, is the lack of resistance, the quasi resignation of those who, when they are not taking the path of rebel dissidence, are supposed to incarnate what separates us from the truth. As if they themselves knew the quasi-ineluctable character of their defeat. The only cry that is sometimes raised is a pitiful "but it's happening too quickly, we are not ready!" as was the case in Belgium with the adoption of children by homosexual couples. Protesting "that cannot be done, that will never be done" for its part provokes a slightly voyeuristic tendency, and the ultimate dishonor is attained when traditionalists are reduced to appealing to arguments of the "psy" type to defend their convictions.

Here it is a matter of thinking on the basis of the fact that far from being a heroic combat, critique seems henceforth to have something redundant about it, as if it merely ratified *something that has already happened,* which has already been carried out, as if it duplicated a prior operation of destruction. Perhaps that is why nothing, or nothing much, grows again where illusion has been destroyed – as if those who pride themselves on having triumphed over it were limited to digging up weeds that are dying or already dead, killed by a ground that is poisoned.

Thus when the never-ending refrain "you believe that this 'really exists', in the sense that it would have the right to impose itself on us, but in fact it is nothing more than a 'social' construction" resounds, no sense of suddenly liberated possibilities makes itself felt. Everything seems to have been said but nothing is produced. The desperately general adjective "social" most often equates with "arbitrary," with what could just have easily been different. Certainly that also signifies that it is now available for change – but what change? And above all, since the nineteenth century, in whose interest is it that nothing resist change? What does the generality that everything is social mean, if not the result of a generalized operation of rendering equivalent? That is to say also the destruction of what mattered in a mode that was irreducible to a generality, of what claimed not an exceptional status but the taking into consideration of its own manner of diverging from the general rule.[2] And is not what is called society then that which is defenseless in relation to the operations of redefinition through the categories of the State and the production of infernal alternatives by capitalism?

I am not denying that the adjective social may have had an eminently positive and constructive sense when the labor movement gave birth to it, in the epoch in which it learned actively, knowingly, to meddle with what was supposed not to concern it, to create relations of cooperation, solidarity, and mutuality, to explore what a "popular" and not a "public" (State) education could mean. But the fact that today critique can end up in the sadness of "it's just a social construction" marks the end of this intensely constructivist moment. The adjective social was emptied out when public order came to rhyme with social peace and the State took in hand and submitted what had been created to its categories. And it is not the supposedly immediately social immaterial labor that will give a positive meaning back to this adjective, which is now honored precisely because it is abstracted from everything that attaches humans, everything that produces relations that aren't interchangeable.

Perhaps one might say that critique, which certainly was a remedy, has become a poison, because it has not known how to defend the truth proper to what is constructed, to what succeeds in holding together and making hold together, what is fabricated

2 "To diverge" must be understood here in the sense that, as in *La Vierge et le neutrino*, I associate it with an ecology of practices taken not as contradictory or incommensurable but as heterogeneous: the manner in which a practice, a way of life, or a being diverges designates what matters to them, and this not in a subjective but a constitutive sense. If they cannot make what matters to them matter they will be mutilated or destroyed.

and yet has the power of a cause, which makes those who fabricated it think, act, and feel.[3] And perhaps it didn't know how to because of its historical anchoring in Science, in the reference to scientific progress substituting a corrosive truth for human beliefs, expelling from the world that which humans, having finally arrived in the age of Reason, no longer had anything to do with. When it celebrates as the progress of reason the destruction of what people are attached to, without accepting that what they are attached to might be what causes people to think, doesn't critique follow the path of Science, discovering a social explanation behind appearances? Even if it meant, in recent decades, itself turning against sciences themselves and discovering that they also could be assimilated to a form of illusion, a social construction like the others.

And certainly there was grist to the critical mill, because scientists have never said nothing but the truth, the whole truth, about what made them practitioners: that was the condition under which their successes could be presented as moral, as representative of the general progress of reason, and also under which the all-terrain judgments that are demanded of Science could be accepted, separating what must be taken into account from what is merely subjective. If at the end of the twentieth century what has been called the "science wars" was able to stage the denunciation, by furious scientists, of the critical reading of the sciences, this was because they were already experts in the matter. They knew that critique dismissed their knowledge as that which they incessantly dismissed as "not scientific," that is, a mere social construct. This war probably belongs to the past, however. With the knowledge economy, critique will be able to

3 This is what Bruno Latour, also struggling against social (de)constructivism, has called "factishes," thus responding to the antifetishism that again and again denounces those who attribute an existence to what is only a construction. See *On the Modern Cult of the Factish Gods* trans. Catherine Porter and Heather MacLean (Durham NC: Duke University Press, 2010) and chap. 4 and 9 of *Pandora's Hope: Essays on the Reality of Science Studies*, trans. Catherine Porter (Cambridge MA: Harvard University Press, 1999)

function in complete redundancy. For new generations of professional researchers, used to the injunction that they must interest industry, the very idea of an achievement imposing criteria of reliability that are more demanding than those of their industrial partners, and of gaining patents, will doubtless appear to be a romantic illusion belonging to the past.

Today the hero of the critical epic has become postmodern. Unlinked from the reference to Science and concluding with the terrible relativism of everything, he resides in a sad hall of mirrors. Emancipation seems to be summed up by the interminable task – which is, apparently, all the more sacred for being interminable – of breaking every reflection, always with the same refrain "it is constructed." That is, unless a new sacred cow is postulated – human rights, democracy – the empty abstraction of which defies critique. How does one critique a postulate? Critique is now in a situation of levitation, something that is, moreover, celebrated by some as the ultimate lucidity, finally assuming the abyssal drama of the human condition. I implore those who may be seduced by the hymn to the death of thinking to consider that there is perhaps a certain obscenity to today's somewhat "chic" radicalism – like a demonstration by the absurd that far from liberating new questions and new possibilities, critique is pursuing the shadow of what had mattered, had caused people to live and think, and is honoring what can no longer cause anyone at all to live or think.

If the question now is that of the causes able to make us think, invent, and act, to allow us to repopulate our devastated history, it is necessary to know a priori that they will all be vulnerable to critical attack, to that which we have carried out like mad chemists systematically submitting everything that they encounter to the acid bath and triumphantly concluding "it doesn't resist!" They will, on the other hand, need the critical, discerning attention that the art of the pharmakon proposes, but then it isn't a question of illusions to defeat, but much rather one of knowing that what can be a remedy is all the more likely to become a poison if it is used imprudently and without experience. And that is a kind of attention that has nothing epic about it, which may have belonged to every epoch and to every tradition.

I recall that I have been addressing myself to those for whom what I proposed might be felt to be a terrible renunciation, the betrayal of what has been most precious to us. But this disarray may be doubled by a cry "that would be to open the door to every kind of monster!" and then the scenario changes, because what is in question with this cry is "the others," those who will be vulnerable to the most monstrous of temptations. Here again, it will be a matter of naming, so as to force thinking. In our so-called modern world, when the hero in the epic genre makes himself into the sworn enemy of the illusions that fetter the process of emancipating humanity, it can have as its consequence the power given to what I will name stupidity.[4]

4 Bêtise is translated here as stupidity; however it is worth noting that this is not unproblematic – stupidity invokes *stupor*, sleep, while "la bêtise," as will be seen here, has nothing passive about it. –Trans.

[1 2]

Stupidity

Just as Gaia cannot be reduced to an object of knowledge, what I will name stupidity cannot be reduced to a type of psychological weakness. It will not be said that "people are stupid" as if it was a matter of some personal defect. Stupidity is something about which it will be said instead *that it seizes hold of certain people.* And in particular it seizes hold of those who feel themselves in a position of responsibility and who then become what I call our guardians.

It is not that those who are responsible for us, those who are found everywhere, at all levels, are stupid, in the sense that it would be sufficient to get rid of them, to take power, to put smart people in their place. And it is not that everyone who is responsible is afflicted with stupidity. The technician responsible for the operation of a network of computers isn't especially, as such. As the saying has it, it is a bad workman who blames his tools, and responsibility, here, implies an attention towards the possible, the capacity to imagine the unforeseen, a learned wariness not towards the situation but towards one's own routines. On the other hand, it is us that those who are responsible for us distrust. While wariness implies a precise situation, procedures to use, and commitments to formalize, those who are responsible for us are defining us as never to be trusted. They are like shepherds who must answer for their foolish flock to whoever entrusted them to look after it.

One might think that by assimilating our guardians, those who I am calling responsible for us, to shepherds who must answer for their flock, I am associating the question of stupidity with what is called "pastoral power," which implies a leader who has received a mandate to assure the safety of those he must guide. Stupidity, however, is rather what remains of this power when there is no longer any mandate, or whenever only an impoverished version of it subsists, staging a recalcitrant humanity, one that is always ready to allow itself to be seduced, to follow the first charlatan to come along, to allow itself to be had by the first demagogue. Those who are responsible for us are not pastors because they are not guiding us towards anything at all; they are in the grip of stupidity because they judge the world in terms of dangerous temptations and seductions that it is a matter of protecting us from.

Today, faced with the intrusion of Gaia, which they can no longer entirely ignore, those who are responsible for us are in suspense, as we are. The "I am aware but all the same..." stance that takes the place of thinking for them is nearly audible, but, in a certain manner we are all in that position. On the other hand, what is not of the order of a common reaction in the face of what is difficult to conceive, of an impotent complaint in the face of what exceeds us, is the reaction – almost a cry – that is typical in the face of certain propositions: "but such a proposition would open the floodgates to..." To hear this cry is to hear what makes the difference between the compassion that is possible for whoever is in a position of responsibility and feels out of their depth, and the distance to be taken with regard to those who I am characterizing as responsible for us. Because this cry is the cry of stupidity.

When "I am aware but all the same" is associated with the cry that invokes the open floodgates, discussion is pointless, because one is dealing not with someone whose reasons would have to be understood, but a being who has been captured, in the grip of something in relation to which any reason will come afterwards, and most often in the mode of the "you should remember that..." Whoever says "you should remember that" is not dumb – those who are seized by stupidity never are. He doesn't plaintively demand to be understood but rather is frightening. Because what makes him react – although perhaps he would like the world to be different, that people be not "like that" – is of the order of a force that one collides with, and, what is more, a force that one feels feeds on all the efforts at persuasion, all the arguments to which one might be tempted to have recourse.[1] Stupidity does not here refer to stupor, to paralysis, or to impotence. Stupidity is active, it feeds on its effects, on the manner in which it dismembers a concrete situation, in which it destroys the capacity for thinking and imagining of those who envisaged ways of doing things differently, leaving them stunned, a stupid and nasty argument may well leave you stunned with the mute perplexity of a "he may be right but all the same," or enraged, which confirms it in turn: you see, with these kinds of people, there's always violence.

1 Naming stupidity is to repeat in a different way the operation that Philippe Pignarre and I attempted in *Capitalist Sorcery* when we named "petites mains" ("minions"), those who are not simply part of the "system" as one says, but who watch over it, who incessantly adjust its articulations, block up its leaks, and extend its hold. In that case too it was a matter of diagnosing a capture and a hold, one which, when the minion says "sorry, but we have to..." makes them say so not as something they suffer but as a commitment that sets them up against those who insinuate that there might be a way of doing or thinking differently. Naming is a risk and one doesn't do just as one pleases with words. The name petites mains was probably associated with too sympathetic an image, as people behind the scenes, who provide invaluable assistance, and the name minions connotes something a little too obsequious, too personal a relation to a powerful person, to permit their use to be changed. We have had to recognize that.

It seems to me that it is necessary today to dare to name the stupidity that seizes hold of those whom capitalism has made endorse the responsibility for maintaining public order. And that even though – and our guardians "are well aware of it, but all the same" – it is systematically and in all irresponsibility activating new sources of disorder. It is not a matter of accusing, as is the case when complicity or corruption is denounced. Such accusations in effect create the idea that if one rid oneself of these sell-outs, everything would be OK – an idea that always favors those who present themselves as saviors, the voice of the people, of the nation... or of the race. And they only reinforce the sense our guardians entertain of the necessity of their mission – their conviction that those who accuse them "don't understand." Those who have been captured by stupidity deserve neither accusation nor indignation. In fact *they deserve nothing* because it is that which they are in the grip of that matters. And what grips them can be sensed at every level of responsibility, and it connects them all, including those who are strangers to the direct interests of contemporary capitalism, including those who have been captured by the pedagogic refrain "what would you do in our place?" and feel themselves responsible for us by proxy.

Gilles Deleuze, from whom I have borrowed this name stupidity, made of it a new problem, one that imposed itself on those who questioned the erring of human thought in the nineteenth century. In his *Alphabet*, when he tackles "H as in History of philosophy" Deleuze carries out a kind of wild gallop. The philosophers of the seventeenth century were, he says, preoccupied with error – how is one to avoid error? But in the eighteenth century, a different problem emerges, that of illusion, of the vulnerability of the mind to the illusions to which it adheres, which it even produces. Then in the nineteenth century, it is stupidity that haunts some, like Nietzsche, Flaubert, or Baudelaire, which fascinates and horrifies them.

That the question of stupidity arises in the nineteenth century doesn't in the least bit signify the discovery of something that previously had been ignored or misrecognized. Stupidity is new, like the coupling of modern States and capitalism is new. It doesn't affect capitalism, because capitalism doesn't fear the opening of the floodgates to anything whatsoever. What it doesn't want, what would be a fettering of the laws of the market, is what would prevent it from being in command when it is a matter of defining the manner in which problems have to be posed. But stupidity does, on the other hand, affect those who view themselves as the inheritors-rentiers of the Enlightenment, those who continue the noble combat against illusions but who – and this makes for a difference that matters – have abandoned its sense of adventure for that of a mission that made them pedagogues. They are those who have to protect others, *those who know, whilst others believe.*

It is a matter here of "thinking by the middle/milieu" in Deleuze's expression, that is to say, without descending to roots nor ascending to the final sense, but grappling with a milieu that is henceforth saturated with multiple versions of the "they believe, we know" that fabricate those who are responsible for us, those who know that behind the floodgates that must not be opened a formidable mass of beliefs are jockeying for position, always ready to invade the stage.[2] In one way or another, the Entrepreneur's demand – that the State ensures the security of his investments – is an ingredient in the matter. But who is the chicken and who is the egg? Couldn't one just as easily say that the State has lent a favorable ear to the Entrepreneur, because what this latter was proposing corresponded to its own sense of its responsibilities, to ensure the orderliness of progress by closing the floodgate to irrational turbulence? In any case, we are grappling with two protagonists who have been coproduced by

2 It is rather remarkable that the cry about the open floodgate is only emitted very rarely when it is a matter of a socio-technical innovation – then one speaks about something satisfying the "needs" of a population, if only to observe later that the supply made a powerful contribution to actualizing the need.

their alliance. Rather than seeking to identify these two protagonists and their respective roles conceptually, "thinking by the middle" here could well signify producing knowledges that conspire to fabricate a different experience of this middle/milieu, to recount our histories differently, *and notably to learn to discern the manner in which stupidity has poisoned them.*

Although it is only an example, it is thus that we have become used to seeing scientists hold that one of their most important and most legitimate missions, in the name of reason, is to hunt down those who they denounce as charlatans, impostors, carrying away a credulous public, a public that is vulnerable to every kind of seduction, susceptible to every kind of irrationality. Recounting the manner in which this role has been taken on, in which it produced the evidence for a scientific reason struggling against opinion, is also to recount the manner in which stupidity has captured the scientific adventure, has contributed to putting the power of proof in the service of public order. This is what I realized when studying the approach initiated by the scientists who in 1784, that is to say, just before the French Revolution, participated in a commission of enquiry nominated by King Louis XVI to investigate the magnetic practices of the Viennese physician, Anton Mesmer.[3]

Around Mesmer's "baquet" (vessel), loaded, according to him, with a curative magnetic fluid, women swooned and the crowd became impassioned, a crowd that was a danger to public order, because it brought magnetism into resonance with affirmation of the equality of humans, all brought into relation by the fluid. The Queen, Marie Antoinette, it was said, was as sensitive to the fluid as the lowliest of her chambermaids. And for the first time, scientists, amongst whom were the masters of experimentation,

3 Léon Chertok and Isabelle Stengers, *A Critique of Psychoanalytic Reason: Hypnosis as a Scientific Problem from Lavoisier to Lacan,* trans. Martha Noel Evans (Stanford CA: Stanford University Press, 1992).

Lavoisier and Franklin, will take on the role of those who "feel themselves responsible" and apply themselves to finding the means to destroy the claims of this charlatan. To do this, they will invent a new type of proof: it will not be a matter of successfully giving a reliable interpretation, one that resists objections, to the crises, and the cures that Mesmer attributed to the fluid, but of devoting themselves solely to the question "does Mesmer's fluid really exist?" For them this signifies: does it have any effects independently of the imagination, in this instance, independently of the knowledge that one has of being magnetized? In other words, the commissioners were there in order to disqualify, and that is what they would do thanks to a series of stagings at the center of which was the power to dupe. Subjects are duped with the complicity of a magnetizer, and the conclusion follows: the imagination can produce the effects that are attributed to the fluid, whereas without the imagination, the fluid has no effect. Mesmer is thus nothing but a charlatan.

Of course the effects felt by those who were duped, who thought that they were being magnetized, didn't have all that much to do with the curative effects that could be observed around the baquet. One ought really to write that the imagination *must be able to explain* such effects, something that has nothing to do with experimental proof. What is more, the members of the committee didn't define the power of the imagination, nor did they envisage the hypothesis that in order to be effective, the fluid demands the imagination. These objections, and many others, were made at the time, but in vain, because it wasn't Mesmer's therapeutic practices that were at stake: what the commission had selected was his claims to give the fluid the power to explain the cures that he obtained. It was these claims that allowed his practice to be submitted to a test imposed in a unilateral manner, a test that resembles experimentation but doesn't seek experimental achievement, just the power to judge. The approach inaugurated by the commissioners, which has illustrated the critical spirit proper to Science ever since, is made to kill things off, to mark a stopping point in a history that is judged to be irrational. And it repeats itself every time that, faced with what he judges apt to raise an unwelcome interest, a scientist concludes "it must be possible to explain that by..." Explaining is no longer a rare achievement but a judgment that manifests the power of reason to dissipate illusion.

Accepting power's offer, placing their science in the service of public order, the commissioners were aware that an abyss separated the event of experimental proof, when a phenomenon has become able to explain itself, from the "it must be possible to explain that by...," which, as it happens, always explains by a general cause, emptying what it explains of any interest. They knew this but to acknowledge it would have been to open the floodgates to the crowd that stuck blindly to the authority of Mesmer the illusionist. Let us not be mistaken though: what the commissioners produced really is of the order of an invention, but having agreed to count themselves amongst our guardians what they invented was the power to dismember a concrete question – what occurs around Mesmer's baquet – in the name of Science, that is to say, to redefine this question in terms of categories that authorize them to conclude: "Move on, there's nothing to be seen here."

One can see why it is so important to emphasize that the hold of stupidity doesn't make those who are vulnerable to it "stupid" because they feel themselves responsible. Those who are made stupid, or dumb, are rather those who are seen as threatening the public order. When one says of a remark that it is "stupid and nasty" one is characterizing something that is remarkably effective, but of a destructive efficacy, producing a paralysis in the thought of whoever it targets. To render the power of stupidity perceptible is thus not just about making perceptible the manner in which it anesthetizes those who it seizes hold of, prohibiting them from wondering, hesitating about the way a situation demands to be approached, felt, and thought. It is also about rendering perceptible the manner in which it commands them to invent the means to subject such situations to unilateral requirements that have the nasty power to dismember them. Because what matters for them is not the situation itself but what is rumbling behind the floodgates, the formidable and formless mass of illusions that only ask to profit from this situation in order to rush on stage.

[1 3]

Learning

The commissioners could have replied: "But what would you have done in our position?" To which just one response stands out: "We aren't in your position." Not a very polite answer, but a salubrious one. To refuse to put oneself in their position is, in effect, to refuse the anonymity that those who feel themselves responsible claim. It is this kind of response that is appropriate when, for example, it appears that the refusal of GMOs puts our guardians in a difficult situation in relation to the rules of the WTO: "if you have given up on the possibility of prohibiting the cultivation of GMOs in European countries, if you now have to be accountable to your masters in the WTO, you have done so without any mandate" – just as it is without any mandate today that those who are responsible for us attempt to impose a freedom of "exchange" on African countries and a submission to IP rights that would be totally ruinous for these countries. And it is without any mandate that they have defined the limits of political action by reference to their necessary subjection to what they call the laws of the market. Concretely, this signifies that what capitalism is now able to make them do includes the task of ensuring our subjection. How are we to put ourselves in their position if amongst the illusions jostling for position behind the floodgates that must be kept shut is now to be found the idea that trying to think the collective future is a legitimate right?

Naming stupidity in order to make it perceptible, in order to make it felt that agreeing to imagine oneself in "the position of..." is to expose oneself to its grasp, is all the more important today given that it is a matter of resisting appeals to unity in the face of the challenge of global warming. Naming stupidity is not a good thing in itself, however. The art of the pharmakon is required. As a remedy, the operation can certainly be demoralizing for our guardians, those who, in order to feel good, need us to put ourselves in their position, *that is to say, allow ourselves to be infected by the stupidity that has captured them.* But every remedy is susceptible to becoming a poison. If the refusal of GMOs was an event, it was not just because the disarray of our guardians had become perceptible, but also because on this occasion, minor knowledges were able to make themselves heard and conspired to fabricate a very different problematic landscape. The floodgates were effectively opened but onto the multiplicity of questions that the order-words "agriculture must be modernized" had silenced. Beyond the generalities correlating the empire of GMOs, which is nothing other than that of industrial agriculture, with a series of quasi-programmed catastrophes, there is no generality that would define a different agriculture, one that is able to compose itself with Gaia, but also to stop poisoning the concrete Earth and its inhabitants, and this whilst feeding ever growing numbers of humans. Not that this is impossible but the possibilities have to be formulated on a case by case, region by region basis, and above all in a mode that confers a crucial place to the knowledges of interested people. The poison here would to underestimate this challenge, the need to learn what it requires, here too on a case by case basis, without postulating a generalized goodwill. Multiple connections are to be created and maintained, never to be considered acquired once and for all.

If we return to the Mesmer affair that allowed me to illustrate the theme of stupidity, the situation is the same. In the minority in the commission, the naturalist Jussieu had called for a renewal and careful study of what he called the traditional "treatment by touch," to which he thought Mesmer's magnetism belonged despite its revolutionary claims. Heeding Jussieu's appeal, studying the traditional therapeutic practices of healers from the countryside, rather than subjecting them to the criteria of judges who were indifferent, even hostile, who in any case had decided to make the right of scrutiny and control of Science prevail, might have meant *to learn (how) to work with healers*. That is to say, with practitioners without formal qualifications who, unlike Mesmer, would not have presented themselves as discoverers, but much more often as the custodians of a transmitted knowledge or gift. And to do that, it would have been necessary not to make the grand break between "those who believe" and "those who know" prevail, and to recognize the healers as those whom it was a matter of learning from and with. Legend has it that Galileo had the courage to murmur "and yet it moves" when he was condemned to recant. But those who condemned him were not his scientific colleagues. To affirm apropos of practitioners with troubling references "and yet they heal" in the face of scandalized colleagues demands much greater courage, the type of courage that researchers not only do not cultivate but that they are actively encouraged to refrain from ("that would open the floodgates to...")

Thus there really is something that is pressing against the floodgates, which it is the task of our guardians to keep shut, a whole mass of learning to do, which is sometimes shocking, always difficult, because it cannot be reduced to the generalities of good sense. It is this mass I am thinking of when I refer to the manner in which Gilles Deleuze characterized the difference between left and right – a difference in nature, he emphasized, not of conviction. This difference of nature refers to the relationship with the State power, and it is why the parties said to be on the left do not cease to betray it: the left needs, in a vital manner, people to think, that is to say also to imagine, to feel, to formulate their own questions and their own demands, to determine the unknowns of their own situation.[1]

The institutions of the State can only disappoint such a need. I will limit myself here to an example from education, when it is dominated by the State imperative of "control and verification," ensuring that whoever has passed a stage is capable of providing comparable answers to the same questions, of responding to the same demands. What such verifications produce is well known. Far from being a simple element of the educational apparatus, they constitute its heart and soul: verification captures what comes before it and defines the transmission of knowledge (an unjustly criticized expression) as a passage from a supposed ignorance to a knowledge that is defined by its conditions of verifiability.[2] This makes school, officially placed under the sign of equality, a systematic producer of inequalities, inequalities that are, in addition, ratified by those interested in maintaining them. One need only think of the sad demand for the equality of opportunity. What does such a demand signify, if not the abstraction of a "whoever" who aims to get themselves recognized as belonging to the set of all those who find themselves offered the same opportunities as all the others, a little like a lottery ticket that has the same chances of winning as all the others? Except that the demand can have as its correlate a returning of the responsibility for one's fate back to whoever didn't seize the opportunities they were offered.

Today, the difference in nature between learning to pose one's own questions and submitting to questions that come from elsewhere is taking on a formidably concrete signification: the possibility of a response to Gaia that is not barbaric could

1 Gilles Deleuze, *Negotiations,* trans. Martin Joughin (New York: Columbia University Press, 2003), 127.

2 Julie Roux, *Inévitablement (après l'école)* (Paris: La fabrique éditions, 2007).

indeed depend on it. Because the responses to give will not be responses to questions that are ready made because addressed to "whoever." They will always be local responses, not in the sense that local means "small" but in the sense that it is opposed to "general" or "consensual." As for the "people," whose thinking is needed in a vital manner, they are never the others, those unreliable, vulnerable to irrationality others about whom our guardians talk without ever including themselves amongst them. Learning to think, to pose one's own questions, to situate oneself by escaping from the evidence of the whoevers, is never permanently acquired, defining an elite against the suggestible flock. The only thing that can be acquired is, rather, the taste and the trust for it. And those who acquire it today know they were lucky and can often recount the encounter or the event to which they owe this experience, the possibility of which school and the media had left them unaware of: not "I think" but "something makes me think."

Learning to recognize and to name stupidity, then, matters, but it is not an end in itself. Rather it is a matter of a condition for something else, an active diagnosis bearing on our milieus, milieus that don't make the learning of this experience that Deleuze calls thinking impossible but exceptional (for the elite, not for "people"). It is an eminently political diagnosis, because it is in these milieus that one also deals with those who are engaged in experimenting with what "thinking" means to live or survive, thinking *in the sense that matters politically, that is to say, in the collective sense,* with one another, through one another, around a situation that has become a "common cause" *that makes people think*. It is a matter of diagnosing the unhealthy character of the milieus in which such experiments will always risk being dismembered, subjected to control and scrutiny, and to regulations that are blind to their consequences, summonsed to provide accounts that are not theirs, destroyed. But also, should this happen, unduly glorified as "the" solution by those who will hasten to condemn them if they do not live up to the promises they have been made to bear. In the world that is ours, one must, of course, be mistrustful of one's enemies, but also of one's (critical) friends, who are always ready to be "disappointed." And yet it is also a matter of trusting that if the occasion is appropriately constructed, people can become capable of acquiring or reclaiming the taste for thinking. That is to say, of discovering that what disgusted them or what they had no taste for, felt incapable of doing, wasn't thinking at all (which is indissociable from a concrete, practical experience) but the indeed rather loathsome exercising of a theoretical abstraction which demands that what one knows and lives be dismissed as anecdote.

But this is a utopia, it will be objected! To accept this objection is to condemn us to barbarism. And it is barbarism to which we are also condemned by the tales and reasoning that we are drowning in, which illustrate or take as a given the passivity of people, their demand for ready-made solutions, their tendency to follow the first demagogue to come along. Is it any surprise, since this is precisely what the hold of stupidity allows and propagates. We have a desperate need for *other stories,* not fairy tales in which everything is possible for the pure of heart, courageous souls, or the reuniting of goodwills, but stories recounting how situations can be transformed when thinking they can be, achieved together by those who undergo them. Not stories about morals but "technical" stories about this kind of achievement, about the kinds of traps that each had to escape, constraints the importance of which had to be recognized. In short, histories that bear on thinking together as a work to be done. And we need these histories to affirm their plurality, because it is not a matter of constructing a model but of a practical experiment. Because it is not a matter of converting us but of repopulating the devastated desert of our imaginations.

The accusation of utopia rests not on the rarity of cases but on that of the narratives, or instead on their "exoticization." Thus in order to affirm that there is nothing to be learned from nonmodern practices of gathering around divisive subjects, it suffices to qualify the unity of societies in which such practices are cultivated as "organic" (closed, stable, based on adherence to self-evident common values, etc.). The case is closed then: to take an interest in such practices would be to pursue an illusory, or worse still, a regressive, ideal.

We know that within our so-called modern societies, however, modes of gathering that stimulate the capacity to do what people are reputedly incapable of doing exist. Without even mentioning scientific practices, when they are alive and demanding, let us take, for example, the manner in which, although nothing prepares them for it, citizens selected at random become capable of participating effectively in juries in court cases, their attentive presence preventing the usual connivance between professionals and their "we will all agree that..." subtexts. It is hardly surprising that professionals periodically dream about working without juries. They evoke the incompetence of mere citizens, but what really worries them is that the role such citizens take on brings uncertainty into the process. We should rather record that the role they have without deserving it because of their merits or competences has the power to stimulate capacities to think, object, and formulate questions that are precisely what is denied when it is said "people aren't capable." And the experience of citizen juries that meet with regard to technico-industrial innovations gives the same type of empowerment, when the procedure is not rigged, that is to say, organized around a ready-made question, or run by communications professionals, whose techniques are addressed, as always, to groups that are supposed to be incapable of functioning without any "framing."

In these two cases, persons who are "anybodies" demonstrate that they are able to learn how to orient themselves in a situation that is complicated and conflictual because the protagonists in this situation are constrained by their presence to produce it in a mode that allows them to take a position, because the apparatus of the meeting has allowed this situation to be "dramatized," unfolded in all its divergent, undecided and conflicting components. In the case of citizen juries the dramatization is all the more remarkable because it is not a matter of a recapitulation of what has been produced in a legal inquiry: the jury carries out the inquiry itself, forcing a confrontation with experts who, in general, do not speak with each other, it unfolds questions that these experts usually forget, taking an interest in consequences that have been ignored or disqualified or externalized, that is to say, reputedly concern other protagonists, who are not onstage.

It will not be surprising that in this world of ours the institution of citizen juries can only have an extremely limited scope, and that forms of public consultation, which are very fashionable, have in fact most often been reduced to cosmetic operations deprived of any consequence. Entrepreneurs demand that the accounts that they must give of their action – if they cannot be avoided – must be predetermined. We've seen all this. How could they accept an institution in which people produce open accounts and learn to interrogate the manner in which problems are formatted, that is to say too the distribution presiding over the formatting: what the State allows capitalism to do, and what capitalism makes the State do. But it is precisely because it is a matter of an institution in which this distribution is liable to lose any consensual evidence that citizen juries matter. Not only because this institution has the capacity of making perceptible the stupidity of those who present themselves before such juries as responsible, the arrogance, the naïvety, the blindness of certain experts, but above all because of what it is, or what it could be – productive of narratives that give those who hear them the taste for what has produced them. Yes, a situation can become interesting, worthy of making people think, able to stimulate a taste for thinking, if it has been produced by a concrete learning process, in which the difficulties, the hesitations, the choices and errors are as much a part of the narrative as the successes and the conclusions arrived at.

[1 4]

Operators

Let's not fool ourselves: if we do not pay attention, the prospect opened up by the example of juries, whether in a court case or as a citizen jury, could bring us back to what it was a matter of avoiding: the contrast between blind and obtuse experts and professionals, and a group of goodwilled citizens who would provide the proof that when the occasion arises, "people" are able to think. To stop at such a contrast would lead to head-on opposition to our guardians and their allies, most notably those who will multiply all the examples that in their opinion testify to the voluntary servitude proving that people will follow the first demagogue to come along, etc.

Head-on opposition is a temptation to be avoided, because it empties out the world, only allowing two virulently opposed camps, which function in reference to one another, to subsist. In so doing, it feeds stupidity, because it accepts the question of knowing "whether or not the people are capable of..." This is the kind of abstract question that leads nowhere, except perhaps to school and its operations of verification: let's see if they are capable.

For my part, I have never encountered "people," only ever persons and groups, and always in circumstances that are not simply a context but which are operative. Thus, what interests me with the example of juries, whether in a court case or a citizen jury, is not that they would manifest the equality of humans when it is a matter of thinking. It is *the efficacy of an apparatus that brings about a "making equal."* Continuing with the contrast with school, it is significant that the efficacy of the apparatus of the jury depends upon the deliberate exclusion of everything that might repeat a situation of a school type, in which one is supposedly ignorant, necessarily needing to be taught something before being authorized to think, always dependent upon those who supposedly know more.

It is crucial to underline that there is nothing demagogical about not presupposing ignorance. Avoiding the repetition of a situation of the school type, that is to say also of avoiding reviving the "I don't understand" that is produced at school, is part of the apparatus in a positive sense. It positively takes into account that having to undertake a course in the knowledge mobilized by genetic engineering before discussing GMO *will never put this innovation in a position of being thought.* The questions that matter always come afterwards, and this afterwards, when at last it comes, will not have been prepared for by the pedagogical exposition, but rather will have been captured. GMO will first have been presented as a consequence of the progress in knowledge, and the difference between GMO in research (carefully disinfected, because things have to be made simple) and GMO from Monsanto will only be evoked in the last instance, if at all.

Avoiding situations that produce inequality is not enough, just as most of the so-called egalitarian modes of functioning, those that make equality into an abstract injunction, claiming to make a clean slate of all the processes that have always already transformed differences into inequalities, are not. Thus in meetings in which "everyone has a right to express themselves": boredom, self-censorship, effects of terror, feelings of impotence in the face of those with big mouths and other unrepentant windbags, questions that get bogged down incessantly in personality clashes or rivalries between people, the gnawing desire that someone "take things in hand," the progressive rout, weary, fragile compromises...it is pointless elaborating this, as it is a shared experience.

If citizen juries are able to escape this poison, like juries in courts, it seems that it is to the extent that the apparatus gathers its participants around a common cause, that is to say, *achieves the transformation of a problematic situation into a cause for collective thinking*. But this cause that makes participants equal cannot be equality itself, or any other cause supposed to transcend particularities and demand equal submission. Equality is a pharmakon too, one that can become a poison when it is associated not with a production but with an imperative, and an imperative that always sanctions its privileged spokespersons. A common cause, endowed with the power to put those it gathers together in a situation of equality, cannot have a spokesperson. Rather, it is of the order of a question, the response to which depends on those it gathers together, which cannot be appropriated by any one amongst them. Or, more precisely, it is a question *the answer to which will be messed up if one amongst those it gathers together appropriates it*.

When the event of an achievement occurs, it is the "questioning"[1] situation that *produces* equality, that is to say, the capacity of "simple citizens" to participate in juries. It is this situation that transmutes what is presented as an expert response, with an authoritative status, into a contribution the importance of which must be evaluated as well as what it makes matter, what it leaves indeterminate. So, woe betide an authoritative expert,

1 To be distinguished radically, of course, from the problem situations dear to the pedagogue, which are defined in terms of the potential learning of pupils, in terms of mental operations that they will have to put to work.

caught red-handed, judging problems about which he has no expertise to be unimportant, something that must be accepted as the inevitable price of progress. It is because they are brought together by a questioning situation that citizen juries can be formidable machines for making experts stutter, or for evaluating the reliability of the expertise on which what is proposed to them rests.

Today, one could say that the intrusion of Gaia produces a questioning situation of this type, calling into question all our stories, our positions, those that reassure, those that promise, those that criticize. The power of the situation is nothing if it isn't actualized in concrete apparatuses however, apparatuses that gather concerned people around concrete situations. The only generality, here again, is of a pharmacological order. We have a need, a terrible need, to experiment with such apparatuses, to learn what they require, to recount their successes, failures and drift. And this culture of the apparatus can only be constructed in real time, with real questions, not in protected experimental places, because what has also to be learned is precisely what such places, because they are protected, take shortcuts on: how is one to hold up in a milieu that is at one and the same time poisoned by stupidity and turned into a hunting ground for the predators of free enterprise? And how is one to do so without closing up on oneself, with fabricating a nice little world that may well become a stakeholder, protecting its particular success in contempt for everyone else (just do what we do!)

That the milieu of a group experimenting with the possibility of a collective regime of thinking and action can at the same time be what poisons it, what threatens it and that to which links have to be created, indicates clearly that any shortcut in thinking here is lethal, and notably any search for a guarantee, but also every transformation of what is experimented with into a model. The questions that such a group raises, because they form part of this group's milieu, *are operative questions*, even and especially if they pretend to be neutral, the questions that judges or voyeurs ask. As for the responses, they will never be general, they will always be linked to the invention of practical means for making a response.

Let us take a fairly crucial example, that of trust, as much the trust between members of a group as between this group and its milieus. Making of trust an operative question is to make two linked senses of this word diverge – let us call them the "having" and the "fabrication" of trust. When the trust that one had turns out to have been misplaced, one feels betrayed, duped, disappointed, disgusted, indignant, but it is powerlessness that dominates, and it can be translated by recoiling, vindictiveness, ressentiment: "I won't get fooled again!" This is in effect what often happens and it is what testifies to the unhealthy character of our milieus: not only can a group be betrayed by those it thought were its allies, but it can be denounced for betraying the trust of those who had celebrated it as exemplary. On the other hand, American activists practicing nonviolent direct action have given us the example of veritable, artful fabrications of trust. What is presupposed here is that betrayal is what everyone will be incited to do during an action. These activists in effect know that what they must prepare for is a test: not only will the police provoke them into violence but the consequences of the action – prosecution, prison, heavy fines – will be designed in such a way as to divide them, to provoke disagreement and mutual accusation. Typically among those who will be selected for prosecution many will feel that they have been put in a situation that they were not capable of dealing with, or that they have been taken hostage by a decision making process that exceeded them and the price of which they now have to pay, or that they have been left hanging at the moment they have to face consequences: shame, ressentiment, disappointment, guilt.

Fabricating trust, for these activists, corresponds to apparatuses that make for *the envisaging of action on the basis of these tests and these foreseeable traps*. And this, once again, implies resisting the fiction of equality, in this instance, everyone's equal capacity to stick to their commitments, to demonstrate a responsible autonomy. It is, on the contrary, a matter of *conferring on the tests to come the power to make the participants feel, think, and dare to speak* in a mode that renders perceptible and legitimate the heterogeneity of everyone's modes of commitment, and what they feel capable of. In short, an entire pragmatics, not of avowal but of imagination and of the creation of the means to make equality pass via differences that are not the object of any judgment, but which will be that which the vectors of betrayal will profit from if they are not taken into account.[2]

Nothing is guaranteed, as is always the case with the pharmacological art. The transformation that confers on the test the power to make think, which constitutes it as an integral part of the questioning situation, however, is able to "treat" what are foreseeable poisoning operations. Attention no longer bears on persons but on modes of collective functioning that in and of themselves render some vulnerable, their possible betrayal being subsequently taken as a reference so as to accentuate the mistrust, intensify suspicion, and thus to anticipate and provoke new betrayals.

Obviously the art of apparatuses doesn't concern stakeholders. These play themselves off against each other in every possible way, but they cannot betray each other as they are united on the basis of the valorization of their respective interests and have no other cause to serve. Nor does it concern those united by the power of a mobilizing cause characterized as a response, a truth with the power to make people agree. Because such a cause communicates with an ideal of homogeneity, where all are mobilized equally by what gathers them together, by what is good in itself. The art of apparatuses is a pharmacological art because those whom it concerns are gathered together by what is, in the first place, a question that requires an apprenticeship. The fabrication

2 Starhawk, *Webs of Power: Notes from the Global Uprising* (Gabriola Island, BC: New Society Publishers, 2002).

of trust is a part of this apprenticeship not only because the possibility of betrayal is taken as a constitutive dimension of the situation, but also because it gives a positive signification to the heterogeneity of the gathering together, through its response. It constitutes this heterogeneity as something that must be recognized, indeed even as something that must be actively produced, a production that requires apprenticeship.

And so perhaps "everyone together," but the ensemble will only be robust or pertinent if what composes the "everyone" is not subject to the "same," a same that refers the responsibility of this ensemble to that which is struggled against. To be reliable, the ensemble must not presuppose a postulated equality, but must translate operations for the *production of equality* amongst its participants. This signifies that it must be of the order of an alloying of heterogeneous elements, not a fusion. What it is a matter of learning, in each case, is the manner of making divergences exist, of naming and taking them into account where otherwise the poison of unspoken, shameful differences would have acted, with its potential for the divisive maneuvers that will inevitably occur. And it is a matter of learning not only so as to resist these maneuvers, but because far from being assimilable to a defect, the production of equality between participants, which demands that their heterogeneity be activated, is also that thanks to which the different dimensions of the situation that unites them will be unfolded.

[15]

Artifices

It should be unnecessary to emphasize that making divergences present and important has nothing to do with respect for differences of opinion, it must be said. It is the situation that, via the divergent knowledges it activates, gains the power to cause those who gather around it to think and hesitate together. I would go so far as to say that the achievement of an alloying, of a practice of the heterogeneous, doesn't require a respect for differences but an *honoring of divergences*. "I respect your difference [of opinion]" is a rather empty thing to say, which smells of tolerance and commits whoever says it to nothing. On the other hand, what can enter into communication with the word "honor" is something that will be apprehended *not as a particularity of the other, but as what the other makes matter*, what makes him or her think and feel, and which I cannot dream of reducing to the "same" without being insulting – the dream is transformed into a nightmare. Because what is thereby grasped, as something that is irreducible to psychology or to a notion as general as culture, is that which, if it is destroyed, would make our world all the poorer. Divergence doesn't belong to a person, rather it is that which makes an aspect of this world matter.

Naming Gaia, naming stupidity, and, now, honoring divergences in so far as they are related to the situation and not to persons, are propositions whose truth derives from their efficacy. An efficacy that one might say is against nature, if one holds to the usual opposition between the natural and the artificial. But with this qualification: that this opposition has no positive sense. The desperate search for that which, being "natural" would supposedly have no need of any artifice, refers in fact, once more and as ever, to the hatred of the pharmakon, of that whose use implies an art.

The natural, in its sadly predictable sense, is what serves as an argument for those who feel themselves to be responsible. Thus many scientists will affirm that people must trust Science, because if they took a measure of everything that scientists don't know, the completely natural reaction would be to relate what these scientists know to opinions like everyone else, opinions that can be ignored if they are disturbing, if they are an obstacle to a rational solution. Similarly those who mistrust user associations worry that these users obey a selfishness that is completely natural, and will, in a sadly predictable manner, call into question those who prevent them from fully enjoying what it is that they use, including mechanisms for the solidarity and protection of workers that it took so many struggles to create.

If the intrusion of Gaia signifies the necessity of learning to pay attention, of accepting inconvenient truths, we are in desperate need of artifices, because we desperately need to resist the sadly predictable. *It is barbarism that is today sadly predictable.* But the test here is once again to abandon with neither nostalgia nor disenchantment the epic style and its grand narrative of emancipation, in which Man learns to think by himself, without needing any artificial prostheses any longer. This grand narrative has poisoned us, not because it would have lured us with the illusory prospect of human emancipation, but because it has given a debased version of this emancipation, one marked by a scorn for those peoples and civilizations that our categories judged well before we undertook to bring them, with their consent or by force, our enlightenment. Do we not recognize ourselves in their rituals, their beliefs, their fetishes, these artificial prostheses that we have been able to free ourselves from?

That the definition given to emancipation has been marked by polemics is nothing to be surprised about, because in our regions it has been associated with struggle. But what has made us a danger to the planet, ready to recognize illusions everywhere, is the way that emancipation has come to coincide with the struggle against human illusions. What has turned sciences into servants of the public order is the way they have defined their achievements, which are primarily creations, the production of prostheses of a new kind, in terms of the denials that they inflict on opinion. Of course, some will propose that illusions be tolerated, but with the gentle scorn of those who think that they have no need of that. The path from scorn to stupidity is completely traced out.

How many times I have felt this scorn when I have described the artifices invented by the American activists. How many times I have heard the sniggers, assimilating their inventions to tricks well known to social psychologists, using catch-all categories, like the performative character of language or symbolic efficacy. These blunt demurrals, which are analogous to the commission's verdict against Mesmer, attributing the action of his fluid to the imagination, are terribly effective. Let us not be mistaken: these really are naming operations, but the efficacy of these operations is the inverse of the efficacy that I am aiming at when I name. In their case, the operation can be phrased: "Move on, there is nothing to think about here." This reminds us that like every efficacious operation, naming is both a remedy and a poison, but also signals that if we do not perceive the poison, if we confuse the name with a category of a scientific, or neutral, type, this is because we are intoxicated. How are we to think without becoming addicted to critical demystification? How are we to deprive ourselves of the gentle poison of the "we have not been fooled, we possess the categories that identify what it is that others put to work without knowing it."

Those who have been poisoned are also those who scorn what I have called the art of the pharmakon, with the same protest as always: what is of the order of the truth requires no artifice to impose itself. Or with the same objection: if the efficacy of a proposition requires an art of cultivation, is not the door open to relativism? What a horrible possibility! Must one not postulate that certain propositions have the power of imposing themselves by themselves, if we want to avoid the conflict of opinions and the arbitrariness of relations of force becoming an unavoidable horizon? The objection is all the more curious for coming from scientists who nonetheless know very well that a scientific interpretation can never impose itself without artifice, without experimental fabrications, the invention of which impassions them much more than "the truth."

And the height of scorn and derision is reached when an analogy between certain artifices and the techniques used in businesses can be denounced: "and why not bungee jumping whilst we're at it, as it works with executives..." And yes, businesses seize hold of everything they can use, with perfect indifference to our sniggers. Sniggers that are emitted in a quasi-automatic manner by those who always again place themselves in the position of the brains of humanity.

Let's not kid ourselves, what provokes the sniggering has a great deal to do with the idea that thought is what is merited, demanding renunciation and solitude. That is why a good number of these "brains" can, on the other hand, bow with respect before the passion of Antonin Artaud, who yelled and screamed that thought is not "in the head." What matters to them is that yelling and screaming exemplify a radical experience, in as great proximity as possible to madness. Artaud then, this consecrated cultural hero, offers us the confirmation of what Man is capable of confronting, at the risk of losing himself in it, the abyss of chaos that must be kept at a distance in order to think. What provokes sniggering is the use of artifices that could be called democratic, those that it is so easy to dismiss as superstitions, or to role-playing games or autosuggestion. What is more, they are artifices that demand a collective, experimental art, radically denuded of any tragic connotation. That the human adventure might pass via the pragmatic learning of techniques that our sniggerers have been so proud to do without seems quasi -indecent, a sort of deliberately infantilizing business.

It is often said that techniques are neutral, that everything depends on their utilization. Substitute for the term utilization the term use and the sense of neutrality changes. It is no longer what allows responsibility to be shifted onto the utilizer, but is what requires precautions, experience, and the mode of attention that every pharmakon demands. The hatred of artifices, always associated with the threat of relativism, is the hatred of the pharmakon. If everything depends on an artifice or an art, then one can make people think anything and everything.

It is automatically evident that one can associate artifices with the worst (grand Nazi rituals, etc.). But is that not precisely why practicing the art of the artifice matters, why we need to cultivate a capacity to discriminate between their uses, an experience of their potential? It has been necessary for me to understand the power of stupidity to understand why the danger could serve as an argument, to understand why those who feel themselves responsible demand that the only legitimate means for political action be those that are guaranteed to be without risk, like children's toys. And for as long as that is what they demand, as long as they are haunted by the threat of a fantasized populace that is always ready to follow the first deliberate agitator, the equality that they dream of will remain an incantation, nullified by the position that they occupy as responsible.

In fact, there is often very little needed between recognizing and ignoring the importance of artifice. Thus Jacques Rancière has described superbly the importance of the old Athenian apparatus, which carried out the choice of magistrates through a lottery.[1] To be sure, this only concerned those who could claim such functions (most notably not women, slaves, or foreigners), but the lottery matters for Rancière because it signifies that those whom a power is conferred on did not conquer it, did not have to beat others, and would not owe their position to a recognition of their merits. He isn't what I have called a whoever, however, as he will have to think to ask questions, to participate in a deliberation. On the other hand, he is an "anyone." Anyone can! And it is as such that he becomes a magistrate. For Rancière, this anyone designates politics as that which supposes and effectuates a disjunction with the natural order – it is natural that the best, or the most competent, or the most highly motivated, govern. But he does not linger on the *efficacy* of the lottery as artifice, an artifice that also characterizes citizen and criminal juries. Those who are selected by a lottery *know* that they are an anyone, and that is doubtless what protects them from the complicity that is so easily established between experts and guardians, those who feel themselves responsible. As anyones they do not owe their role to some merit that would distinguish them, *and this role, as a result, obliges them*, constrains them to look for what the situation demands, and not to think themselves capable of defining it. Certainly chance loses its conceptual imperiousness as a pure signifier of politics. But it engages a thinking of efficacy that it is a matter of learning to honor.

Chance is then all the more interesting as it situates very precisely the efficacy of artifice. It is not a matter of allowing chance to

1 Jacques Rancière, *Hatred of Democracy*, trans. Steve Corcoran (London: Verso, 2013).

decide, but of having recourse to a procedure that, between us and what we do, makes what is not ours exist, opening up a situation in relation to which we do not have to claim to be up to it. The manner in which the idea of appealing to chance can shock, when it is a serious matter, where the elected one should be selected, demonstrates just how far merit and motivation as reasons have created a void around themselves, to the point of dismissing as arbitrary everything that cannot present such reasons. But chance is also the simplest of artifices. One day, perhaps, we will experience a certain shame and great sadness at having dismissed the age-old traditions – from the auguries of antiquity to those of seers, Tarot readers or cowrie shell diviners – as superstition. Then we will know how to respect their efficacy, independently of any belief, the manner in which they transform the relationship of those who practice them to their knowledges, in which they render them capable of an attention to the world and its scarcely perceptible signs, which open these knowledges up to their own unknowns. On that day, we will also have learned just how arrogant and careless we have been in regarding ourselves as not needing such artifices.

[16]

Honoring

Ticklish Gaia, such as I have named her here, cannot be associated with either prayer, which is addressed to divinities able to hear us, or with the submissiveness that this other blind divinity, honored under the name "the laws of the market," demands. To honor Gaia is not to hear a message that comes from any kind of transcendence, nor is it to resign ourselves to a future under the sign of repentance, that is to say, the acceptance of a form of collective culpability – "we must accept that we must change our way of life." We haven't chosen this way of life, and all the knowledgeable sociological narratives that tell us about the modern individual tell us about a "remainder," about what remains when what had the power to cause us to think, feel, and act together has been destroyed, when free enterprise has conquered the right not to pay attention, and has shunted the burden of risk management onto the State.

If it is a matter of honoring Gaia, one must not repeat in her regard what were perhaps the errors of Marx's inheritors: fabricating a point of view organized around a humanist version of salvation, in which the question posed communicated directly with the emancipation of humankind that is finally capable of overcoming what separates it from its truth. Perhaps it is a question of salvation, but in the sense that this reference doesn't guarantee anything, authorizes nothing, is not associated with any "and so," and doesn't communicate with any providential morality reducing the intrusion of Gaia to that which our history needed in order to be fully accomplished. Responding to Gaia's intrusion by means of triumphalist slogans/order-words that put the ends of humanity on stage would always show that we have learned nothing, again and as always, accepting the grand epic narrative that makes us, always us, into the pathfinders. Didn't we invent the fateful concept of humanity? It is, instead, a matter of detoxifying the narratives that have made us forget that the earth was not ours, in the service of our history, narratives that are everywhere, in the heads of all those who in one manner or another feel themselves responsible, the bearers of a compass, the representatives of a direction that must be maintained.

It is not enough to denounce the pastors, responsible for a herd that they must protect from seduction and illusion. If I have offered a eulogy to artifice, it is because it is necessary for us to reclaim, to reappropriate, to relearn that whose destruction has turned us into a quasi herd. And what I have called artifice translates this necessity. We who are the inheritors of a destruction, the children of those who, being expropriated of their commons, have been the prey not only of exploitation but also of the abstractions that made them into whoevers, we have to experiment with what is likely to recreate – to take root again as one says of a plant – or to regenerate the capacity to think and act together.

I haven't stopped emphasizing that such experimentation is political, because it is not a question of making things better, but of experimenting in a milieu that is known to be saturated with traps, infernal alternatives, and impossibilities concocted as much by the State as by capitalism. But political struggle, here, doesn't happen through operations of representation but much rather through the production of repercussions, through the constitution of "resonance chambers" such that what happens to one group makes others think and act, but also such that where one group achieves something, what they learn, what they make exist, becomes so many resources and experimental possibilities for others. *However precarious or small it might be* each achievement matters. None will suffice to appease Gaia, but all will contribute to responding to the trials that are coming, in a mode that is not barbaric.

Of course, it is not a matter of substituting a culture of experimental achievements for the necessities of open political struggle, which is all the more necessary for having to invest spaces that are reputedly beyond politics, in which experts are activated, calculating limits, attempting to articulate the measures to be taken, with the imperious necessity of durable growth. Even the apparently sensible notion of limit is bearer of the threat of the sad but determined "we must..." that announces barbarism. Limits are what are negotiated between our guardians, they are imposed on the herd, and leave in the shadows the fact that in our world, riven by radical inequalities, a veritable miracle would be needed for limits not to be a factor in even greater inequality. And that would be the case whatever the prodigious accomplishments of the technique that announces to us today that Man will become capable of manipulating matter atom by atom, of shattering his biological limitations, of beating old age and of living in intelligent houses that will satisfy his slightest desires.

Political struggle should happen everywhere that a future that none dare imagine is being fabricated, not limiting itself to the defense of acquired gains or the denunciation of scandals, but seizing hold of the very question of this fabrication. Who pays the technicians, how are scientists educated, what promises make the wheels of fascination turn round, to what dreams of the rich is one entrusting the issue of restarting the economy? Scientists and technologists themselves need such questions to be posed, and some – such as Jacques Testart – have the courage and lucidity to ask that they are, that political struggle move in on technoscientific innovation, where at the moment apolitical slogans of the kind "the planet is in danger, let us save research!" resonate. But it is precisely because political struggle must move in everywhere that it cannot be thought of just in terms of a victory or a conquest of power. The point here is not moral but pragmatic: no power, from wherever it comes, however legitimate it may be, can as such produce the responses Gaia's intrusion obliges, at all levels.

The GMO event offers an example of a coupling of a new type between anticapitalist struggle (and Monsanto is a fairly precise figure for this capitalism that concocts a barbaric future) and the production of thought. Those who are responsible for us have got to promising second- (or third-) generation GMOs, with the slogan "if you want the marvelous others that will follow, you have first to accept this one." But by doing this they raise even more questions. They have not managed to isolate the antiGMO activists, to label them ecoterrorists, because knowledges have been produced that have publicly left the experts stammering, because the biotechnologies that produce patents can no longer rally their scientific colleagues quite so easily in a grand crusade against the rising tide of irrationality, because certain of these colleagues have been led to ask themselves questions at the same time as the public. To be sure it is rare for geneticists to betray genetics from the inside, like Christian Velot did, that is to say, to put their research grants, and so their careers, at risk, so that what their colleagues won't talk about is made known. But the GMO event is one of those events (one thinks also of struggles over the question of medication, or now over energy) that, if appropriately "activated," can help scientists call their role into question – as much the role that is assigned to them in the knowledge economy as that which has for much longer put them in the clutches of stupidity, making them the guardians of the moral order, of rationality against an opinion which, as Bachelard put it, is always wrong. The outline of a possible new kind of researcher, inventing the means for independence in relation to their sources of finance, which enslave their practices, is the order of the day. This possibility is part of the stakes that couple political struggle and creation, because whatever happens we will need scientists and technicians.

What is missing in the GMO event? Firstly a political resonance chamber that is up to the job: even political allies, when their electoral credibility is what matters to them, are frightened of getting every dimension of the event communicated, and notably politicizing the question of progress that technoscientific rationality bears, or that of the knowledge economy, its patents and partnerships. "More research money is needed" is a theme that still works and is worth trying, as is "the French say no to GMO," the spineless reprising of a refusal that is often reduced to a matter of opinion polls and the respect of public opinion (even if it is wrong). But perhaps what is also missing is its having been celebrated as an event, its having been named such, its having generated witnesses who learn to recount what they owe to it, what it has taught them, how it united them, how it forced them to learn from one another. We need, we desperately need, to fabricate such witnesses, such narratives, such celebrations. And above all we need what such witnesses, narratives, and celebrations can make happen: the experience that signals the achievement of new connections between politics and an experimental, always experimental, production of a new capacity to act and to think. This experience is what I, after Spinoza and many others, will call joy.

Joy, Spinoza writes, is that which translates an increase in the power of acting, that is to say too, of thinking and imagining, and it has something to do with a knowledge, but with a knowledge that is not of a theoretical order, because it does not in the first place designate an object, but the very mode of existence of whoever becomes capable of it. Joy, one could say, is the signature of the event par excellence, the production or discovery of a new degree of freedom, conferring a supplementary dimension on life, thereby modifying the relations between dimensions that are already inhabited – the joy of the first step, even if it is uneasy. And joy also has an epidemic potential. That is what so many of the anonymous participants, like me, tasted in May 1968, before those who were to become our guardians, the spokespersons of abstract imperatives, dedicated themselves to have us forget the event. Joy is not transmitted from the knowledgeable to the ignorant, but in a mode that itself produces equality, the joy of thinking and imagining together, with others, thanks to others. Joy is what makes me bet on a future in which the response to Gaia would not be the sadness of degrowth but that which the conscientious objectors to economic growth have already invented, when they discover together the dimensions of life that have been anesthetized, massacred, and dishonored in the name of a progress that is reduced today to the imperative of economic growth. Perhaps, finally, joy is what can demoralize those who are responsible for us, bringing them to abandon their sadly heroic posture, and betray what has captured them.

No one is saying that everything will then turn out well, because Gaia offended is blind to our histories. Perhaps we won't be able to avoid terrible ordeals. But it depends on us, and that is where our response to Gaia can be situated, in learning to experiment with the apparatuses that make us capable of surviving these ordeals without sinking into barbarism, in creating what nourishes trust where panicked impotence threatens. This response, that she will not hear, confers on her intrusion the strength of an appeal to lives that are worth living.